3rd International Conference on Energy Materials and Applications (ICEMA 2018)

IOP Conference Series: Materials Science and Engineering
Volume 446

Salamanca, Spain
9 – 11 May 2018

ISBN: 978-1-5108-7997-3
ISSN: 1757-8981

Printed from e-media with permission by:

Curran Associates, Inc.
57 Morehouse Lane
Red Hook, NY 12571

Some format issues inherent in the e-media version may also appear in this print version.

Printed by Curran Associates, Inc. (2019)

For permission requests, please contact the Institute of Physics
at the address below.

Institute of Physics
Dirac House, Temple Back
Bristol BS1 6BE UK

Phone: 44 1 17 929 7481
Fax: 44 1 17 920 0979

techtracking@iop.org

Additional copies of this publication are available from:

Curran Associates, Inc.
57 Morehouse Lane
Red Hook, NY 12571 USA
Phone: 845-758-0400
Fax: 845-758-2633
Email: curran@proceedings.com
Web: www.proceedings.com

Table of contents

Volume 446

2018 3rd International Conference on Energy Materials and Applications

9–11 May 2018, University of Salamanca, Salamanca

Accepted papers received: 22 October 2018
Published online: 28 December 2018

Preface

Papers

Functional material design and analysis

IOP Conf. Series: Materials Science and Engineering **446** (2018) 011001 doi:10.1088/1757-899X/446/1/011001

PREFACE

The 2018 3rd International Conference on Energy Materials and Applications (ICEMA 2018) was held in Campus Viriato of University of Salamanca, Zamora, Spain on May 9-11, 2018. It was dedicated to discuss and resolve issues related to Energy Materials and Applications. The major goal of this conference is to bring together academic scientists, engineers, industry researchers in order to exchange and share their experiences and research results, and discuss the practical challenges encountered and the solutions adopted.

The proceedings present a selection from papers submitted to the conference from universities, research institutes and industries. 16 papers were accepted from 40 presentations including submissions from 15 countries and regions: Spain, Singapore, Korea, Turkey, China, Egypt, Malaysia, Germany, Pakistan, Ecuador, Italy, Morocco, Mexico, Taiwan and Colombia, providing research in Materials Science and Engineering, Construction and Environmental Engineering, Clean Energy and New Energy Technology, Electrical Engineering and Solar Energy Utilization. All of the papers were subjected to peer-review by conference committee members and international reviewers. The selection of papers was based on the quality and relevance to the conference. Here we would like to thank all the technical program committee members who made great efforts on paper reviewing.

This conference program is also highlighted by 2 keynote speakers, 2 plenary speakers and 2 invited speakers to share their research works with us, and they are Prof. Jesús Toribio, University of Salamanca, Spain; Prof. David T.W. Lin, National University of Tainan, Taiwan; Assoc. Prof. Jaroslaw Milewski, Warsaw University of Technology, Poland; Prof. Nezihe Ayas, Anadolu University, Turkey; Assoc. Prof. Qi Li, Tsinghua University, China and Prof. Mohammed Garoum, University Mohammed V in Rabat, Morocco.

Since 2018 is the 800th anniversary of the founding of the University of Salamanca, the conference was held in Campus Viriato of that University. Salamanca is the capital of the province with the same name located in western Spain. It is a world-renowned university city with the oldest university in Spain and one of the oldest in Europe, which is the famous University of Salamanca. In order to know more about the city and have more time to communicate with each other, we held an excursion on May 11th.

Last but not least, our appreciation also goes to the chairs, the technical program committee members, the invited speakers, and the conference secretary, who made great effort to make the conference run smoothly.

Next year, the conference will be held in Beijing, China, and we look forward to seeing you there!

Conference Chair
Prof. Dr. Jesús Toribio, University of Salamanca, Spain

Peer review statement

All papers published in this volume of *IOP Conference Series: Materials Science and Engineering* have been peer reviewed through processes administered by the proceedings Editors. Reviews were conducted by expert referees to the professional and scientific standards expected of a proceedings journal published by IOP Publishing.

IOP Conf. Series: Materials Science and Engineering **446** (2018) 012001 doi:10.1088/1757-899X/446/1/012001

Effect of the temperature variation on the hardness and microstructure of TiSiNO coatings obtained by sputtering

L García-González[1], F López-Huerta[2], D J Araujo-Pérez[1], L Zamora-Peredo[1], N. Flores-Ramírez[3], S. R. Vásquez García[4], T. Hernández Quiroz[1] and J Hernández-Torres[1]

[1] Centro de Investigación en Micro y Nanotecnología, Universidad Veracruzana, Boca del Río, Veracruz 94294, México
[2] Facultad de Ingeniería, Universidad Veracruzana, Boca del Río, Veracruz 94294, México
[3] Facultad de Ingeniería en Tecnología de la Madera, Universidad Michoacana de San Nicolás de Hidalgo, Morelia, Michoacán 58030, México
[4] Posgrado de Ingeniería Química, Universidad Michoacana de San Nicolás de Hidalgo, Morelia, Michoacán 58000, México
E-mail: leagarcia@uv.mx

Abstract. Titanium-Silicon Oxynitride (TiSiNO) coatings were synthetized by reactive co-sputtering technique on stainless steel 304 substrates varying the substrate temperature during the deposition. Two targets were used simultaneously, the first one of Ti on a DC source at 100 W and the second was a Ti-Si-O target on a pulsing RF source at 180 W. X-ray diffraction identified three crystalline phases: titanium nitride (TiN), titanium oxide (TiO) and silicon oxide (SiO_2), where the SiO_2 phase presented compressive stresses at low temperatures and tension stresses at higher temperatures, along with a change in the preferential orientation. The coating obtained at 200 °C showed the higher hardness value of 14.8 GPa, attributed to a lower grain size and high compressive stress. Finally, all coatings presented high fracture toughness, since the was a lack of any type of crack for all applied load values.

1. Introduction

In recent years, the research to develop hard materials has been increased, mainly because the mechanical applications as cutting tools, aerospace and industrial applications are very attractive [1]. Generally, a single hard material could present problems due to their low fracture toughness, however, to combine two or more materials as a coating, can provide enhanced mechanical properties without losing the original properties of the substrate.

From the vast amount of materials that exist to form coatings, a lot of attention has been paid to binary ceramic materials like the titanium nitride (TiN) and silicon oxides (SiO_2), but it has been reported that composite coatings could improve considerably the mechanical properties [1], [2], [3], some of the most important composite quaternary systems are the Ti-Al-N-O or the Ti-Si-N-O, where the Ti-Si-N-O reached values up to 32 GPa [4]. Nevertheless, the high hardness values in quaternary systems are not easy to reach, since most of the Ti-Si-N-O works report values between 5 and 6 GPa [5]. Therefore, to find a path to reach high hardness values of these coatings could be very important. Moreover, there is a small amount of works regarding this matter, owing the difficulty to produce these high-quality quaternary systems is hard.

Between the different techniques to fabricate these coatings are the chemical vapor deposition (CVD) and physical vapor deposition (PVD), where the PVD are well accepted because they provide a better control of the coating growth and it does not generate a lot of contaminant waste. The sputtering is a very versatile PVD technique, that allows the deposition of metallic or ceramic materials at the same time, having the possibility to control the growth of composite materials. In this sense, some works report the growth of composite quaternary systems using the sputtering, however, most of these works try to control the growth by making changes of the atmosphere inside the chamber during the deposition, whether varying the gases flow [6] or altering the power of the sources during the process [4]; moreover, most of these works use two reactive gases, nitrogen and oxygen, during the sputtering deposition process; additionally, the effect of the substrate conditions, like the temperature, hasn´t been studied. The employment of temperature during the deposition on binary systems has had a positive effect, because they promote a better atom movement and diffusion on the substrate to form better crystalline phases [7].

For the above, in this work, quaternary TiSiNO coatings are synthetized using the Sputtering technique, without the use of oxygen as reactive gas, varying the substrate temperature during the deposition process. The effect of the temperature of the coatings in its microstructure and hardness are studied by X-ray diffraction (XRD) and Vickers microhardness respectively.

2. Methodology

2.1. Substrate preparation and coating synthesis

Steel substrates 304 were used, with dimension of 2 in × 1 in. The steel plates where then polished to obtain a mirror finish with chromium oxide. Finally, the steel was cleaned using ultrasonic baths of liquid soap, a mixture of xylene ethanol and acetone in (1:1:1) proportion and a pure ethanol bath, the samples were dried with nitrogen and placed inside the sputtering chamber afterwards. A sputtering system intercovamex V3 was used for the growth of the coatings, where two targets were used, the first target was a pure Ti target (99.95%) placed in a direct current source (DC) with 100 W, while the other target was a Ti-Si-O target (99.98%) on a radiofrequency source (RF) with 180 W. Two gases were used, an inert Ar gas set at 18 sccm and a reactive nitrogen gas set a 2 sccm, leading to a working pressure of 3.5×10^{-2} Torr. The substrate temperature during the deposition was varied by 100, 150, 200, 250 and 300 °C. All samples were growth until their thickness reached 1μm. This configuration allowed the dispense of the oxygen gas during the deposition process.

2.2. Characterization techniques

The microstructure, hardness and fracture toughness were studied on the TiSiNO coatings with the X-ray diffraction (XRD) and Vickers microhardness. The XRD measurements were made on a Bruker D8 advance diffractometer using a graze angle configuration, with a Göbel mirror and a 1° incident angle, step size of 0.03° and 1 s step time in a 2θ swept from 35 to 95°. Further XRD analysis were performed using the PDF-02 database from the ICDD [8], additionally, the Scherrer formula was used to compute the grain size of the coatings [9]. The hardness was studied in a Vickers indentator Mitutoyo H-125, a load swept was made from 500 grf to 0.5 grf. With the complete load swept it was possible to obtain the real hardness value of the TiSiNO coatings, by eliminating the contribution of the substrate. This was performed using the Korsunsky model [10] according to equation 1.

$$Hv_c = Hv_s + \frac{Hv_f - Hv_s}{1+\kappa\beta^2} \quad (1)$$

Where H_{vf} is the hardness of the coating, H_{vs} is the hardness of the substrate and H_{vc} is the hardness of the composite substrate-coating. k is a dimensionless parameter that relates the material response to the indentation and the coating thickness, and β is the relative indentation depth.

3. Results and discussion

3.1. Microstructure

The X-ray diffractograms of the TiSiNO coatings are shown in Figure 1.a, there, it is possible to observe the formation of well-defined peaks, which correspond to crystalline phases that could be identified in the database as a silicon oxide (SiO_2), titanium oxide (TiO) and titanium nitride (TiN), according to the cards PDF 01-070-2540, PDF 01-073-9742 and PDF 00-002-1159, respectively. It is possible to appreciate that the most intense peak is formed at about ~ 43.6° in 2θ, that correspond to the plane (011) of the SiO_2 phase, the same phase also shows another well-defined but lesser intense peak at about ~ 50.7° corresponding to the plane (200). The second most intense peak is located at ~ 74.7° that is related to the TiN phase along with the peak at ~ 90.6°. Finally, the less intense peaks at ~44.5 and ~90.9° correspond to the TiO phase. Due to the higher intensity of the SiO_2 and TiN phases, they could be related to a better phase formation and can be referred as the primary phases, which -as a whole- they form the quaternary system Ti-Si-N-O. It is important to notice that there is the formation of two titanium phases against only one from silicon, this could happen because during the deposition process two targets that contained Ti were used, and it is very likely that the nitrogen gas had better affinity to the metallic Ti target than the ceramic Ti-Si-O target, where the SiO_2 was detached.

Due to the great crystalline quality seen, the coatings form very intense and narrow peaks, which makes it impossible to distinguish changes on Figure 1.a, however, in Figure 1.b and 1.c a closer look to the peaks of the SiO_2 phase show that there are some changes within the SiO_2 phase, neither the TiO nor the TiN phases are explained since it was harder to identify visible changes. From Figure 1.b it is possible to observe that the coatings with the low deposition temperature are more intense on the (011) direction than those with the higher temperature, in the other hand, in the (200) direction, Figure 1.c, the coatings with higher temperature are more intense than the coatings with low temperature, from here, we could say that the tend to change their crystal growing direction on the SiO_2 phase, this could be explained since the TiN and TiO phases prefer a growth in the (200) direction, so when the SiO_2 rearranges due to the extra diffusion caused by the temperature, it will try to grow as the other coexisting phases within the coating.

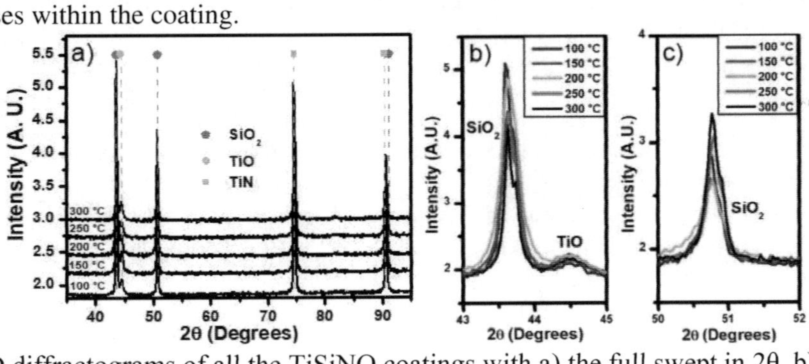

Figure 1: XRD diffractograms of all the TiSiNO coatings with a) the full swept in 2θ, b) a close-up to the (011) SiO_2 peak, and c) an approach to the (200) SiO_2 peak.

With the XRD information it was possible to obtain additional data on the microstructure of the coatings, i.e. the grain size and the percentage of the phases in the coating, which are reported in Table 1. Firstly, it can be observed that there is a change in the peak center, this behavior can give us good information about the stress of the coating. According to the SiO_2 card the center is at 43.622°, and the coatings with low temperature are at the right of that value, indicating the presence of compressive stress, while the samples at higher temperature are left from that value indicating a rather tensile stress. Additionally, the intensity change was corroborated for the SiO_2 phase, where the peak related to the (011) tends to decrease at 250 and 300 °C.

The coatings grain sizes are also reported in Table 1 for each of the phases, we can observe that the grain sizes of the samples tend to decrease as the deposition temperature increases, however, this effect occurs until a certain temperature, then the grain sizes tend to increase; for the SiO_2 phase case the grain size decreases from 46 nm at 100 °C to 35 nm at 200 °C, then at 250 °C and 300 °C the phase grow up to 45.5 nm and 50.8 nm respectively, the same occurs with the TiN phase where at 200 °C reached the lower grain size value of 26.8 nm, then grow again up to 40.9 nm at 300 °C. It is well known that quaternary samples could have a hard time to co-exist due to the different structures they possess [5], [11], so a diffusion and rearrangement process is foster with the application of the temperature, however at low temperatures between 100 and 200 °C a transitory phase could occur before initiating the crystallization process of the sample at higher temperatures.

A similar effect takes place for the TiO phase, nevertheless, the lower grain size value was obtained at 250 °C instead of 200 °C but the amount of this phase is very low as seen in the phase percentage, barely reaching 1% in some cases, so it could be expected that this phase will play a minor role on the properties of the sample. Moreover, the SiO_2 is the mainly phase present in the coatings, but the TiN amount is enough to make it relevant for the properties of the coatings. Additionally, it is possible to observe that the SiO_2 phase amount decreases as the temperature increases, while the TiN phase tends to increase as the temperature increases. This effect could be explained since the TiN phase is formed in a reaction with the N_2 sputtering gas, while the SiO_2 and TiO could just be extracted from the Ti-Si-O target without the need of a reaction, and the temperature applied to the substrate could change the reaction kinetics, favoring the formation of the TiN at higher temperatures.

Table 1: Information of the center (C), intensities (I) as well as gran sizes and percentage of the phases and peaks observed in the XRD measurements.

T °C	SiO_2 peak		TiO peak		TIN peak		Grain size (nm)			Phase percentage (%)		
	C (°)	I	C (°)	I	C (°)	I	SiO_2	TiO	TiN	SiO_2	TiO	TiN
100	43.62	164	44.096	6.6	74.67	52.7	46	31.1	31.1	67.1	0.7	32.2
150	43.65	132	44.507	10	74.69	39.7	38.6	28.9	28.5	63.7	1.1	35.2
200	43.66	145	44.476	6.8	74.7	36.3	35	27.8	26.8	68.6	0.9	30.5
250	43.61	124	44.512	7.1	74.72	59.8	45.5	21.6	31.8	60.9	0.6	38.5
300	43.61	113	44.202	9.1	74	70.4	50.8	24.7	40.9	58.6	0.8	40.6

3.2. Hardness

In Figure 2 the adjustment of the Korsunsky model on the Vickers measurements is visible for the coatings, on a plot of the relative indentation depth versus the experimental hardness. The form of an inverted "s" of the plots is well known, since at high load values the substrate harness values is more visible while at lower load values the contribution to the hardness is mostly from the coating, and the logarithm scale for the relative indentation depth helps to show this behavior.

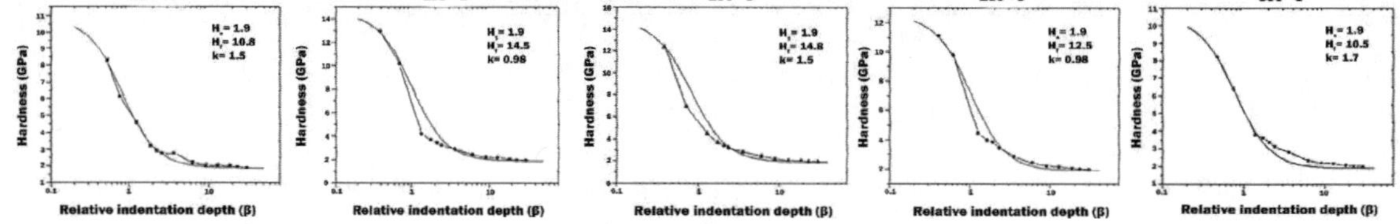

Figure 2: Korsunsky adjustment model for the Vickers hardness values of the Ti-Si-N-O

From the Korsounky model it was possible to despise the hardness contribution from the substrate and to compute the hardness of the coatings, the values of this procedure are reported un Figure 3, where the TiSiNO coatings have hardness values between 10 and 15 GPa, which could be considered as hard coatings. The coatings do not show a linear dependency of the hardness, the coating with a substrate temperature of 100°C reached a 10.8 GPa hardness, and it increases to 14.5 and 14.8 GPa for the coatings at 150 °C and 200 °C respectively, but then the hardness drops to 12.5 and 10.5 GPa for the 250 and 300°C coatings. This behavior can be attributed to the grain sizes seen in the XRD measurements, since it is well known that at lower grain sizes the hardness in higher [12], and the coating at 200 °C had the smaller gain sizes for the SiO_2 and TiN phases, which lead to the highest hardness value of 14.8 GPa. This Grain size - hardness dependency it also demonstrated in Figure 3, where the bigger the grain size is, the lower the hardness, moreover, the hardness values are also corroborated with the compressive stress seen in XRD.

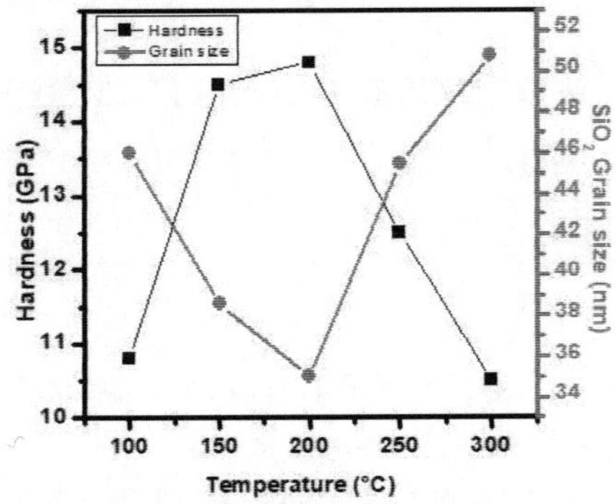

Figure 3: Hardness of the TiSiNO coatings compared to the Grain sizes of the SiO_2 phase.

Finally, it is important to mention that no fracture nor fissures were seen during the measurement of the Vickers hardness, so it could be said that the samples have a high fracture toughness because of the temperature application during the sputtering process, since others works without the temperature application report that the TiSiNO coatings are easily fractured [5].

4. Conclusions

TiSiNO coatings were synthetized by reactive co-sputtering varying the substrate temperature during the deposition process. The temperature favored the formation of highly crystalline coatings where three phases where identified the SiO_2, TiN and TiO, but the amount of the TiO phase was minimal, whereas the SiO_2 and TiN act as the main phases for the sample. The temperature to rearrange the structure to the (200) preferential orientation started at 250°C where the grain sizes started to grow, before that a transition phase occurred where the grain sized decreased and the preferential orientation was (011). The changes in the microstructure caused a change in the hardness of the samples, being the sample at 200 °C the hardest, nevertheless, the temperature favored a better fracture toughness since neither coating presented visible cracks.

References

[1] Y.Y. Chang, *et al.*, 2017 *Surf. Coatings Technol.*, no. June, pp. 1–10.
[2] R. Haubner *et al.*, 2017, *Int. J. Refract. Met. Hard Mater.*, vol. 62, pp. 210–218.
[3] J. K. Park, *et al.*, 2007, *Plasma Process. Polym.*, vol. 4, no. SUPPL.1, pp. 902–905.
[4] L. García González *et al.*, 2013, *J. Mater. Eng. Perform,* vol. 22, no. 8, pp. 2377–2381.
[5] L. García-González *et al.*, 2008, *J. Mater. Eng. Perform.*, vol. 17, no. 4, pp. 580–585.
[6] L. García, *et al.*, 2009, *Inf. Tecnol.*, vol. 20, no. 1, pp. 21–28.
[7] A.A. Ahmed, *et al.*, 2017, *Mater. Sci. Semicond. Process.*, vol. 63, no. February, pp. 137–141.
[8] "The International Centre for Diffraction Data - ICDD - ". http://www.icdd.com/
[9] A. L. Patterson, 1939, *Phys. Rev.*, vol. 56, no. 10, pp. 978–982.
[10] A. M. Korsunsky, M. R. McGurk, S. J. Bull, and T. F. Page, 1998, vol. 99, no. 1–2, pp. 171–183.
[11] J. Sjölén, *et al.*, 2007, *Surf. Coatings Technol.*, vol. 201, no. 14, pp. 6392–6403.
[12] F. Tang *et al.*, 2010, *Scr. Mater.*, vol. 63, no. 2, pp. 192–195.

Acknowledgments

The authors would like to thank the "PFCE 2016-2017" corresponding to the "DES técnica Veracruz".

IOP Conf. Series: Materials Science and Engineering **446** (2018) 012002 doi:10.1088/1757-899X/446/1/012002

The influence of the expanded clay granules ratio on the thermal conductivity and thermal diffusivity of gypsum plaster-based composites's.

Khaoula Amarray[b], Mohammed Garoum[a*], Saad Raefat, Najma Laaroussi and Aziz Ettahir.

Materials, Energy And Acoustics Team (MEAT), Mohammed V University in Rabat - EST Salé, 11060, 227 Avenue Prince Héritier, Morocco.
[a]garoum1@yahoo.fr, bkhaoulaamarray@research.emi.ac.ma

Abstract. Thermophysical properties of building materials such as thermal conductivity and thermal diffusivity are mandatory in any study of thermal performance of buildings. In this work, measurements of these two thermophysical properties were performed on four circular samples prepared by mixing gypsum plaster with different mass proportions of expanded clay granules. The water ratio was kept constant in all mixtures. Then the thermal diffusivity and thermal conductivity were determined using the flash method and the hot plate respectively. Results show that the addition of the expanded clay granules to the gypsum plaster has a negative impact on the thermal properties. This was due to absorption of the mix water and plaster by the clay granulates at the earlier operation of samples preparation.

1. Introduction

Understanding the thermophysical properties of building materials is nowadays considered a necessity in the building sector in order to wisely choose the most technically and economically efficient building. In order to achieve the desired objectives regarding the thermal insulation of buildings, it is advantageous to minimize the cost of construction and decrease its energy bill. The principle of insulation has always been based on the juxtaposition of several insulating materials, which is expensive, occupies a significant volume and enables limited energy efficiency. The use of lightweight aggregates can overcome this problem, due to their favourable thermal characteristics, their porous structure and their mechanic resistance.

Researches related to the improvement of thermophysical characteristics of composite materials based on cement, clay and plaster are unlimited. DAVRAZ et al. [1] studied the effects of unit weight, porosity and grain size on the thermal conductivity of various light aggregates. As for MAALOUFA et al. [2] tempted to improve the thermal properties of clay by adding cork or aggregates of pearlite and make a comparison between the two mixtures. The purpose of BUMANIS et al. [3] study is to identify a relationship between the mechanical and thermal properties of cement mixed with expanded glass aggregates. A thermomechanical study of lightened concrete with cork and olive pomace was made by EL BAKKOURI et al. [4]. A comparison made by CHERKI et al. [5] between the thermal properties of a composite material based on cork and plaster with those of plaster is conducted in order to motivate the use of this mix material as false ceiling. Lamrani et al. [6] investigated the experimental study of thermal properties of a new ecological building material based on peanut shells and plaster. Raefat et al. [7] discussed the influence of adding expanded granular perlite to gypsum plaster thermal diffusivity using a numerical estimation based on the global minimization procedure. In this study, we study the effect of expanded clay granules on the thermophysical characteristics of plaster-based materials.

2. Experience program

2.1. Materials and mixture proportions

This study was conducted to characterize the influence of the incorporation of expanded clay beads Figure 1 on the thermophysical properties of plaster. Before determining the sample formulations we measured the physical properties of the aggregates. These characteristics concern the particle size distribution, the bulk densities as well as the water absorption of aggregates. Each value selected for these different characteristics is derived from an average calculated on three measurement results in order to ensure good and representative results. After classifying the different particle sizes of the expanded clay beads, we obtained three classes (8-10mm), (10-12.5mm), (12.5-16mm). The aggregates of the class (12.5-16mm) are subsequently dried in an oven at 70 °C before making samples for a period of 10 days until the mass becomes constant (the state of constant mass is assumed when the mass does not vary by more than 0.1% after 1 hour of drying). Four cylindrical samples of $100 \times 100 \times 20\ mm^3$, presented in Figure 2, were prepared consisting of gypsum plaster and expanded clay granules. The mixing rate remained the same for all mixtures (0.625), the only variable is the granule's mass fraction, which is 2%, 6% and 10% compared to pure plaster sample of 0%, and these proportions are picked in order to evaluate the influence of expanded clay beads on the thermal properties of plaster. To minimize the moisture content in the pores, the four samples were dried in the oven at 65 ° C for one week after 24 hours of air drying.

Figure 1. View of the class (12.5-16mm) of expanded clay granules

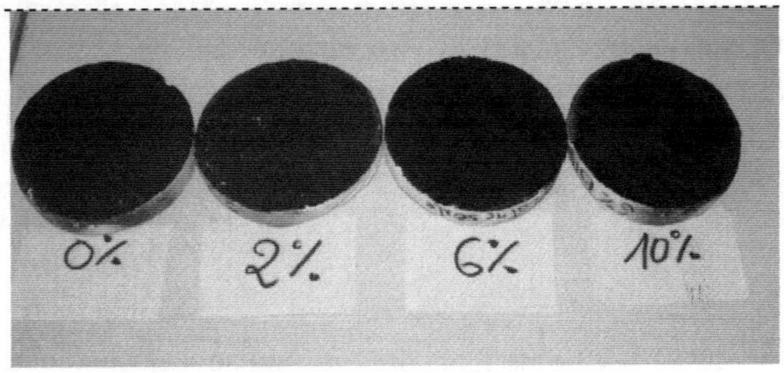

Figure 2. The four samples studied

2.2. Experimental method

2.2.1 Transient Hot plate method

It is the most straightforward method dedicated to thermal conductivity measurement. The method is based on the measurement of the temperature at the centre of a circular heating element $of\ (100 \times 100 \times 0.1 mm^3)$ inserted between two samples [8]. Its electric resistance is $R = 101.51\ \mathbf{\Omega}$. The principle of the method is to apply a voltage U = 12.08V through the heating element interposed between the sample and the polyethylene foam. A thermocouple is glued to the inner face of the heating element. The sample, heating element and insulating foam are placed between two aluminium blocks whose thermal conductivity is too high to bring the system to steady state as quickly as possible [9]. Its schematic is illustrated in Figure 3.

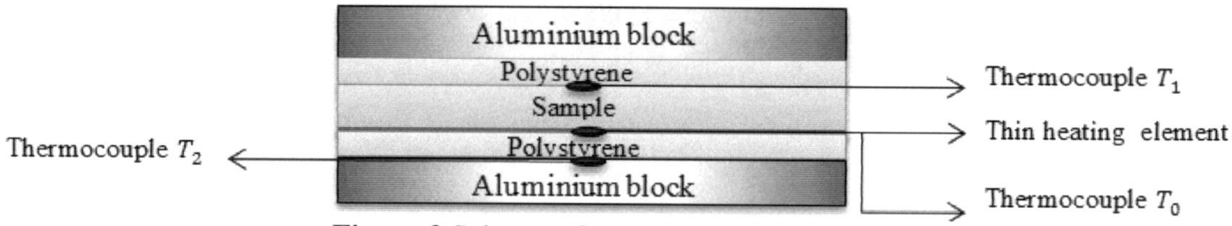

Figure 3.Schema of experimental device

The thermal conductivity of the sample is calculated using the following equation:

$$\lambda = \frac{e_1}{T_0 - T_1}\left[\frac{U^2}{R.S} - \frac{\lambda_2}{e_2}(T_0 - T_2)\right] \tag{1}$$

With e_1 is the sample's thickness, $e_2 = 10\ mm$ is the heating element thickness, $\lambda_2 = 0.047\ W.m^{-1}.K^{-1}$ *is* the thermal conductivity of the heating element *and S* corresponds to its section.

2.3. Flash method

The concept of the method is very basic. As indicated in figure 5 the front face of a sample of $(100 \times 100 \times 20\ mm^3)$ receives a uniform pulse of radiant energy $\Phi_0(W.m^{-2})$ of a short duration τ coming from flash lamp. A thermocouple in contact with the rear face of the sample makes it possible to record the rise in temperature at the moment when the front face receives the pulse. The thermal effusivity value is calculated from the temperature response on the opposite side of the sample. Thermal diffusivity is estimated either by the parker method [10] or by the Degiovanni method [11]. The increase of the temperature difference (T (t) - T (0)) is defined via its Laplace transform indicated by:

$$\theta(p) = \frac{\frac{q}{p}(1 - e^{-\tau p})}{h^2\frac{sonh(ke)}{\lambda k} + 2hcosh(ke) + \lambda k.sinh(ke)} \tag{2}$$

τ is the elapsed time from the flash pulse heating and p stands for the Laplace parameter and $k = \sqrt{\frac{p}{a}}$, a, λ and e are respectively, the thermal diffusivity, the thermal conductivity and thickness of the sample.

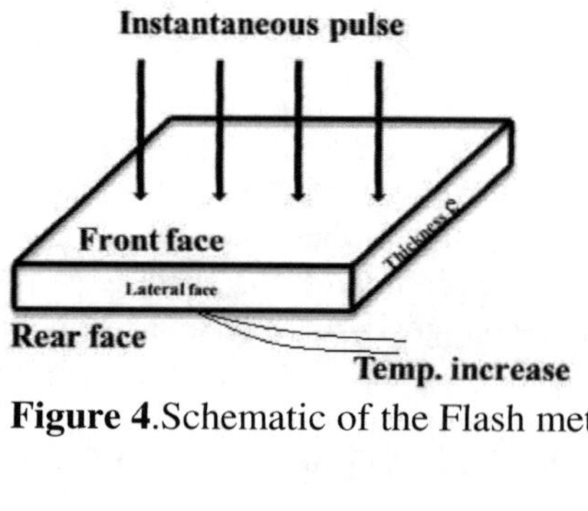

Figure 4.Schematic of the Flash method

3. Results and discussion

3.1. Density

Table 1 groups the results of measurement of the bulk density of the 4 samples according to their volumes and their masses after drying.

Table 1. Samples dimensions and density.

Samples	Thickness (m)	Diameter (m)	Mass (kg)	Density $\rho(kg.m^{-3})$
PA10	2.205	10.37	168.17	928
PA6	2.145	10.05	163.21	933
PA2	1.998	10.10	152.9	955
PA0	1.855	10.05	141.33	965

PA10, PA6 and PA2 represent the plaster samples mixed with different mass fractions of expanded clay granules of 10%, 6% and 2% respectively. While PA0 represents the pure plaster sample used as the reference. From Fig 5, which represents the density as a function of the mass fraction in the four mixtures, we noticed that when the mass fraction of expanded clay granules increases, the bulk density of the sample decreases. The composite's density decreased from 965 (kg/m^3) of PA0 to 928 (kg/m^3) for PA10. In fact following the increase in the fraction of expanded clay granules, the porosity of the plaster increases, this explains the decrease in the density.

3.2. Thermal conductivity

The thermal conductivity of gypsum plaster embedded with expanded clay granules was obtained by the hot plane method in stationary regime. The average values and the measurement deviation were presented on table 2.

According to table 2, we notice a significant increase in thermal conductivity up to 41% from 0.1847 W/m.k for PA0 to 0,315 W/m.k for PA10. This increase could be related to the first quantity of water and plaster absorbed by aggregates during the samples preparation; indeed. The departure of a certain proportion of water and plaster inside the aggregates after being dried leads to the increase of the apparent thermal conductivity of the composite samples with their clay granules mass proportion.

Table 2. Thermal conductivity values of mixtures.

Mixture	Test	The thermal conductivity λ(W/m/K)	Measurement deviation (%)
PA0	1	0.1872	1.34
	2	0.1835	0.67
	3	0.1835	0.67
	Mean value	**0.1847**	
PA2	1	0.2283	4.33
	2	0.2412	1.08
	3	0.2464	3.25
	Mean value	**0.2386**	
PA6	1	0.2857	0.26
	2	0.2863	0.05
	3	0.2873	0.30
	Mean value	**0.286**	
PA10	1	0.320	1.55
	2	0.314	0.44
	3	0.312	1.11
	Mean value	**0.315**	

3.3. Thermal diffusivity

Table 3. Thermal diffusivity values of mixtures.

Mixture	Test	$a\times10^{-7}$(m²/s) Parker	Measurement deviation (%)	$a\times10^{-7}$(m²/s) Degiovanni	Measurement deviation (%)
PA0	1	1.797	3.086	1.722	6.217
	2	1.722	1.216	1.617	0.259
	3	1.777	1.938	1.550	4.391
	4	1.754	0.619	1.542	4.885
	Mean value	**1.7432**		**1.621**	
PA2	1	1.809	2.964	1.827	0.385
	2	1.854	0.550	1.793	1.484
	3	1.894	1.596	1.797	1.264
	4	1.9	1.918	1.863	2.363
	Mean value	**1.864**		**1.82**	
PA6	1	2.25	4.022	2.276	7.498
	2	2.13	1.525	2.091	1.240
	3	2.135	1.294	2.066	2.421
	4	2.137	1.202	2.036	3.838
	Mean value	**2.163**		**2.117**	
PA10	1	2.628	1.399	2.551	0.552
	2	2.529	2.421	2.544	0.276
	3	2.605	0.511	2.455	3.232
	4	2.605	0.511	2.598	2.404
	Mean value	**2.592**		**2.537**	

Results presented in table 3 show that the thermal diffusivity obtained using the flash method of the foor samples studied. We find that for the same class of granules, the thermal diffusivity increased from 1.7432×10^{-7}(m²/s) for PA0 (pure gypsum plaster) to 2.592×10^{-7}(m²/s) with the increase of the proportion of the expanded clay granules in the mixtures. The evolution of the thermal diffusivity as a function of the mass fraction of the expanded clay balls is at the same order of magnitude as the thermal conductivity with an increase of 32%.

4. Conclusion

The present work treated an experimental study on the thermal behaviour of plaster composite materials and expanded clay granules when they are both mixed. Conductivity and thermal diffusivity were measured using the hot plate and flash methods respectively. Knowing the dimensions and the mass of the mixtures, the density was deduced easily. The influence of the incorporation of expanded clay granules on thermophysical properties on a composite material based on plaster was examined. The analysis of the experimental results leads to the conclusion that the diffusivity and the thermal conductivity of the composite material increased, contrary to the decreasing of the density. The increase on the conductivity and the thermal diffusivity is mainly due to the amount of water absorbed by the expanded clay granules. As perspective, in further studies the insulation of clay granules should be treated carefully in order to prevent the clay aggregates absorption.

References

[1] A. E. A. Mertin Davraz, Murat Koru, "The effect of physical properties on thermalconductivity oflightweight aggregate," vol. 15, pp. 85–92, 2015.

[2] Y. Maaloufa, S. Mounir, A. Khabbazi, J. Kettar, and A. Khaldoun, "Thermal Characterization of Materials based on Clay and Granular: Cork or Expanded Perlite," *Energy Procedia*, vol. 74, pp. 1150–1161, 2015.

[3] D. Architektūra, I. R. Statyba, G. Bumanis, D. Bajare, and A. Korjakins, "Mechanical and Thermal Properties of Lightweight Concrete Made from Expanded Glass," vol. 2, no. 2, 2013.

[4] A. El Bakkouri, H. Ezbakhe, T. Ajzoul, and A. EL Bouardi, "ETUDE THERMOMECANIQUE DU BETON ALLEGE AVEC DU LIEGE ET DU BETON ALLEGE AVEC DES GRIGNONS D'OLIVE," no. May, 2016.

[5] D. B. A.Cherki, A.Khabbazi, B.Remy, "Granular cork content dependence of thermal diffusiv thermal conductivity and heat capacity of the compos site material / Gran nular cork bound with plaster .," vol. 42, pp. 83–92, 2013.

[6] M.Lamrani, N.Laaroussi, A.Khabbazi, M.Khalfaoui, M. Garoum and A. Feiz, "Experimental study of thermal properties of a new ecological building material based on peanut shells and plaster," *Case Stud. Constr. Mater.*, 2017.

[7] RAEFAT, Saad, GAROUM, Mohammed, LAAROUSSI, Najma, et al. Thermal diffusivity and adiabatic limit temperature characterization of consolidate granular expanded perlite using the flash method. In : *IOP Conference Series: Materials Science and Engineering*. IOP Publishing, 2017. p. 012004.

[8] Y. Jannot, V. Felix, and a Degiovanni, "A centered hot plate method for measurement of thermal properties of thin insulating materials," *Meas. Sci. Technol.*, vol. 21, no. 3, p. 35106, 2010.

[9] Y. Jannot and B. Remy, "Measurement of thermal conductivity and thermal resistance with a tiny hot plate," no. August 2015, 2010.

[10] W. J. Parker, R. J. Jenkins, C. P. Butler, and G. L. Abbott, "Flash Method of Determining Thermal Diffusivity, Heat Capacity, and Thermal Conductivity," vol. 1679, 1961.

[11] A. Degiovanni, M. Laurent, R. Prost, A. Degiovanni, M. Laurent, and R. P. Mesure, "Mesure automatique de la diffusivité thermique," 1979.

IOP Conf. Series: Materials Science and Engineering **446** (2018) 012003 doi:10.1088/1757-899X/446/1/012003

Logistic regression model for the grooving process of hybrid polyester resin composite materials reinforced with natural fibers and fiberglass

C F Pérez, J G Paredes, R X Valencia, W H Vaca, E P Vásquez

Technical University of Ambato, Faculty of Civil and Mechanical Engineering, Material and manufacturing research group, Av. Los Chasquis and Río Payamino, Ambato, Ecuador.

corresponding author's Email: cf.perez@uta.edu.ec

Abstract. Greencomposites materials are being integrate into the manufacturing of products in diverse fields. For its assembly, are required secondary manufacturing processes such as the machining of slots and holes. In these processes, the greencomposites present great sensitivity to their manufacture. For the study of this concept, an experimental design was developed to analyze the grooving process and verify the effects of the cutting parameters with two types of tools (HSS and HSC) on surface quality (Ra) and the delamination factor (Fd) of the material. An analysis of the experiment was carried out to verify the significant effects of the different factors involved, in addition a multiple linear regression analysis and binomial logistic regression to predict surface quality and delamination in other conditions. The results obtained show the logistic regression model gave a better prediction for the grooving process of the greencomposites, both for the surface roughness and for the delamination factor. Finally, this study concludes that the carbide tool (HSC) has better surface quality with respect to the conventional tool of high speed steel (HSS).

1. Introduction

Studies on natural fiber have generated greater interest for its mechanical characteristics similar to synthetic fibers that meet conditions such as: low density, high resistance to stress and deformation, ease of surface modification and wide availability [1-3]. The manufacture of composite materials reinforced with natural fiber has been made by means of techniques such as: manual conforming with resins, pultrusion, filament winding, vacuum bag molding and resin transfer molding. [4]. As result of these industrial processes, components for vehicles, machines, furniture, etc. are obtained. There are some components that must necessarily be assembled in a system, machine or mechanism; for this reason, machining is an essential option to generate coupling geometries and inherently determine the structural integrity of the compound in these geometries [5-8].

Slot manufacturing is considered the most important machining process in assembly operations. However, there are untoward when evaluating the surface quality due to the non-homogeneity of the fibers and anisotropic characteristics of the material. Mechanical fatigue delamination [7, 9], it is a problem faced by composite materials, resulting in superficial damage to the cutting geometry when machined.

In addition, another defect that occurs in the machining is the surface roughness (Ra), this feature influences the quality of fit, tolerance and friction between elements and therefore the quality of assembly [10, 11], this is unfavorable in the performance of mechanical parts and consequently leads to production costs.

According [4], these two phenomena involve great interest at the moment of being evaluated due to the variation and behavior according to the material that is being experienced and establishes that it is essential to determine the appropriate cutting parameters. According to studies conducted by [12], states that the behavior of load displacement in composite materials reformed with glass fiber / epoxy with circular cuts, concluded that as the size of cut increases, the maximum load of the composite plate decreases. An investigation carried out by [13], studied the residual tensile strength after drilling in fiberglass reinforced epoxy matrix composites and determined the cutting parameters for machining.

There are several studies that use experimental techniques of analysis and optimization; as factorial designs, Plakett-Burman, Taguchi and response surface [10, 14, 15]; all of them have made viable the studies to improve the machining processes. In addition, have identified that the main cutting parameters that affect the surface quality of the machining are between others the cutting speed, feed speed, type of tool and variables such as the properties of the composites. On the other hand, authors like [16, 17], used orthogonal matrices and analysis of variance to determine the level of influence that each factor has on the response variables. Research on the subject referring to the search for probabilistic logistic regression models to predict the best machining conditions in terms of quality, cost and time [18-23].

In the present study, the approach based on the design of experiments (DOE) [24], the relationship focused on the level of dependence and interaction between the different factors (cutting speed, feed speed, tool and fiber orientation) generating a minimum appearance of surface roughness (Ra) and delamination factor (Fd) in green composites. Finally, a logistic regression model is used to obtain an acceptable level of roughness and delamination on grooved surfaces from suitable cutting parameters.

2. Experimental Development

2.1 Preparation of greencomposites

The composite materials for the experimental study have been reinforced with different natural and synthetic fibers: polyester reinforced with cabuya fiber plus glass (CGFRP) and polyester reinforced with abaca fiber plus glass (AGFRP). The configuration of the natural fibers used in the biocomposite presents unidirectional flat tissue. The main properties are described in **table 1**. To elaborate the test pieces, a vacuum stratification process was used according to the following volumetric fractions: 70% (polyester matrix) and 30% (natural fibers plus glass fiber 375 gr/cm^3). The dimensions of the test specimens for the experimental test are (400x250x6mm) with a total volume of 600 cm^3.

Table 1. Properties of composites hybrids [3, 25]

Description	Density (g/cm^3)	Tensile strength, ultimate (MPa)	Young´s modulus (MPa)	Flexural strength (MPa)	Elasticity modulus (MPa)
CGFRP	1,46	80	5272	120	8053
AGFRP	1,48	138	4473	127	5091

2.2 Machines, equipment and tools

For the analysis of grooves in the specimens of composite material has been established to use two types of drills diameter 6mm; the first HSS high speed cutter (SU'S _EM-4144) of 8% Co, right propeller 30 ° and 4 teeth and a HSC carbide drill (CERIN_66SU.060063575) right propeller 10 ° with breaker and 6 teeth. A machining center M-1000 TRAVIS Fagor control was used, that works with speeds of 0 to 8000 rpm. To measure the values of the surface roughness, a SJ-210 Mitutoyo rugosimeter with 1cm probe displacement was used. To evaluate the level of surface delamination, the scanning microscope SEM "Scanning Electron Microscope VEGA3 SEM" was used.

2.3 Design of experiments

The methodology established for the experimental test was a design of experiments (DOE). The two quantitative variables (*Vc*) and (*Vf*) and the categorical variable (*H*) and the orientation factor (parallel PA and perpendicular PE) was evaluated at two levels respectively as detailed in **table 2**, solving a factorial design 2^4. The experimental model consists of four replicates having a total of 64 experimental runs as shown in **table 3**.

Table 2. Assignment of levels to the machining factor

Machining parameters	Low	High
Cutting speed V_c (m/min)	48	60
Feed rate V_f (mm/rev)	0,1	0,15
Tool H	HSS	HSC
Orientation M_o (machined according to orientation of natural fibers)	PA	PE

2.4 Grooving process

The process of grooving in two directions was carried out according to the scheme shown in **figure 1a**, where the combination of machining parameters (*Vc, Vf, H*) is detailed. The depth of the slot is 4mm; being, the depth per pass of the tool of 1 mm. The path of each slot has a length of 100mm.

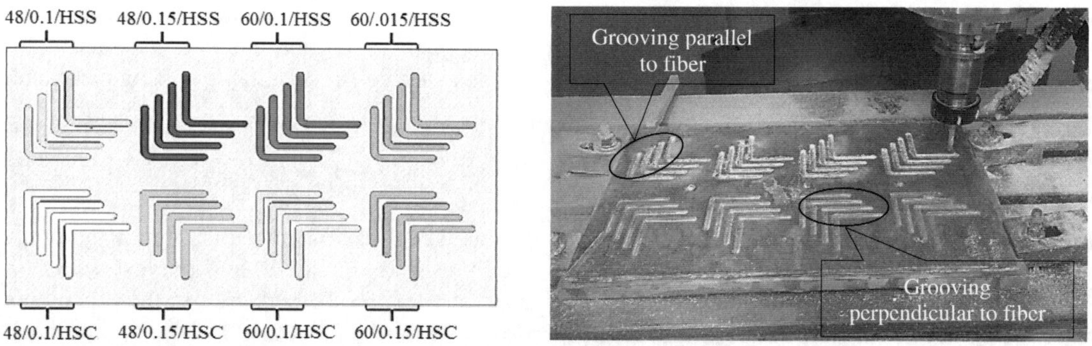

(a) **(b)**

Figure 1. Scheme for the manufacture of grooves (Grooving). a) Combination of machining parameters, b) arrangement of the experiment.

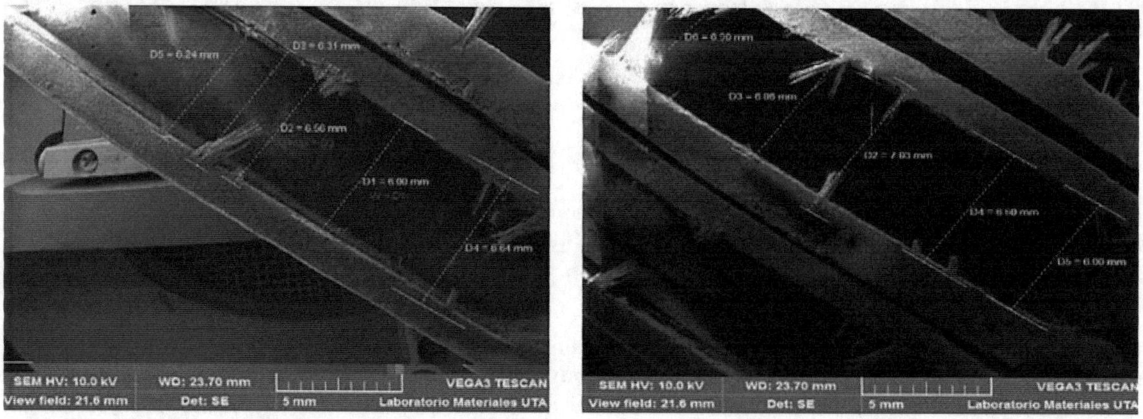

(a) **(b)**

Figure 2. Measurement of the delamination in the two machining directions. a) Parallel, b) perpendicular.

IOP Conf. Series: Materials Science and Engineering **446** (2018) 012003 doi:10.1088/1757-899X/446/1/012003

2.5 Measurement of surface roughness (Ra) and delamination factor (Fd)

Measurements were made on the walls of the grooves in both machining directions. To determine the surface delamination, a SEM scanning microscopy was used, as shown in **figure 2**. Four measurements of delamination by orientation were obtained. The delamination factor (Fd) from the delamination measurement that is calculated by equation 1 [4]; where (W_{max}) is the maximum measure of delamination and (W) is the width of the slot.

$$Fd = W_{\max} / W \tag{1}$$

2.6 Logistic regression analysis

A logistic regression has been used to model in depth the interpretation of the estimators, that is, the experimental test consists of independent variables (Vc, Vf, H, Mo) with two categories or levels also known as dichotomous, and the dependent variable is (Ra and Fd) also dichotomous. The equation of the logic model are presented below:

$$\pi_i = \frac{e^{Z_i}}{1+e^{Z_i}} \tag{2}$$

Where (π_{in}) is the probability of success, (Z_{in}) is the independent variable linearly combined.

$$Z_i = \beta_0 + \beta_1 X_1 + \beta_2 X_2 ... + \beta_n X_n \tag{3}$$

For the resolution of this model, two logical conditions were established to establish a level of surface quality acceptability with a level of significance of 5%. First, Ra> 0.5 "Surface roughness Ra is not acceptable" and Ra <0.5 "Surface roughness Ra is acceptable"; second, Fd> 1.06 "The delamination factor Fd is not acceptable" and Fd <1.06 "The delamination factor Fd is acceptable".

3. Results and Discussion

3.1 Experimental Analysis

The factorial designs have been resolved graphically in Minitab®18.1 Statistical Software. **Table 3** shows the average results of (Ra) and (Fd) of the four replicates obtained. From the analysis of factorial design graphs were obtained that demonstrate the influence of the parameters of cut according to the surface rugosity (Ra) and delamination factor (Fd). **Figure 3**, shows the interaction of the cutting speed as a function of surface roughness; this specifies that as the cutting speed increases from 40 to 60 m/min the surface roughness decreases from 0.7510 to 0.6845 μm for the CGFRP, a Unlike the AGFRP, the surface roughness decreases from 0.6971 to 0.6845 um. On the other hand, the surface roughness of the AGFRP presents a significant variation when a forward speed of 0.10 mm / rev is applied, the roughness decreases from 0.7573 to 0.6245 μm, but, if you work with the same forward speed, the roughness presents a percentage variation of 5.81% in the CGFRP.

Table 3. Average results of Ra and Fd for AGFRP y CGFRP

N°	Vc	Vf	Tool	Orientation	Ra Abacá	Fd Abacá	Ra Cabuya	Fd Cabuya
1	48	0,10	HSS	PA	0,716	1,077	1,126	1,133
2	60	0,10	HSS	PA	0,716	1,186	0,97	1,072
3	48	0,15	HSS	PA	0,804	1,185	1,016	1,103
4	60	0,15	HSS	PA	1,062	1,27	0,791	1,155
5	48	0,10	HSC	PA	0,475	1,129	0,291	1,053
6	60	0,10	HSC	PA	0,297	1,038	0,422	1,01
7	48	0,15	HSC	PA	0,691	1,048	0,245	1,084
8	60	0,15	HSC	PA	0,438	1,038	0,396	1,095
9	48	0,10	HSS	PE	0,78	1,154	1,26	1,083
10	60	0,10	HSS	PE	1,128	1,141	0,843	1,058
11	48	0,15	HSS	PE	1,084	1,068	1,328	1,177

IOP Conf. Series: Materials Science and Engineering **446** (2018) 012003 doi:10.1088/1757-899X/446/1/012003

12	60	0,15	HSS	PE	1,122	1,149	0,723	1,278
13	48	0,10	HSC	PE	0,521	1,007	0,273	1,167
14	60	0,10	HSC	PE	0,371	1,082	0,609	1,157
15	48	0,15	HSC	PE	0,507	1,123	0,47	1,135
16	60	0,15	HSC	PE	0,351	1,023	0,593	1,033

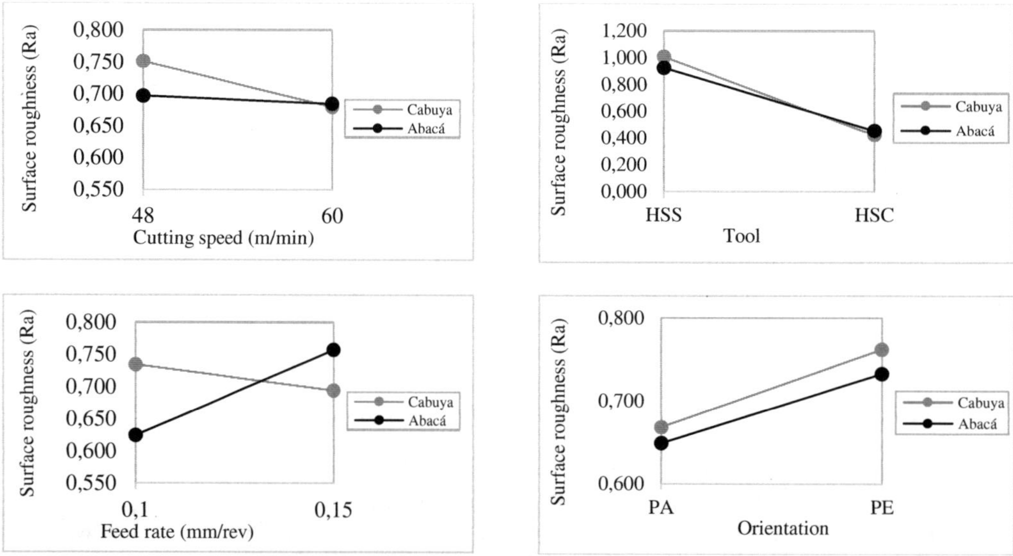

Figure 3. Effects of machining parameters on Ra.

Figure 4 shows that the surface quality is directly influenced by the type of tool, so the use of a tool (HSC) provides a minimum surface roughness and a minimum delamination factor compared with the tool (HSS). In addition, the parallel orientation is influenced at the time of obtaining a minimum surface roughness in the composites; but, when evaluating the delamination factor the AGFRP behaves better when working in a perpendicular direction.

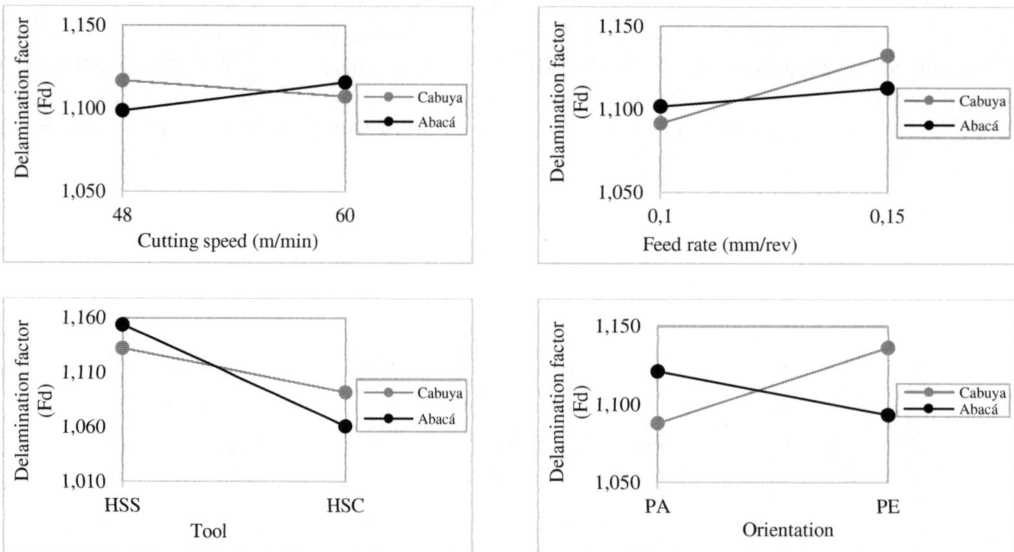

Figure 4. Effects of machining parameters on Fd.

3.2 Binary logarithmic regression analysis

For the analysis the program R "Rstudio" was used, where the dependent variables *(Ra)* and *(Fd)* and the factors of the experiment are modeled and analyzed (x1: cutting speed, x2: Forward speed, x3: tool, x4: orientation). To determine the surface quality of the AGFRP according to the logic model established in **tables 4 and 6**, is specified that the factors cutting speed and tool are significant when evaluating the surface roughness *(Ra.* In the same way, only the tool is influential in the moment of evaluate the delamination factor *(Fd)*.

Table 4. Binary logistic regression, Ra Abacá

	Estimate	Std. Error	z value	Pr(>|z|)
Intercept	-5,07E+03	4,70E+03	-1.078	0.281088
x1	2,99E+02	1,03E+02	2.911	0.003598
x2	-3,13E+04	1,93E+04	-1.617	0.105796
x3	-5,98E+03	1,56E+03	-3.839	0.000124
x4	7,89E-13	8,88E+02	0.000	1.000.000

Table 5. Binary logistic regression, Ra Cabuya

	Estimate	Std. Error	z value	Pr(>|z|)
Intercept	4206154	269242405	0.016	0.9875
x1	-0.21427	0.09319	-2299	0.0215
x2	-1888568	2012579	-0.938	0.3480
x3	-2298759	269241130	-0.009	0.9932
x4	-257127	111824	-2299	0.0215

Table 6. Binary logistic regression, Fd Abacá

	Estimate	Std. Error	z value	Pr(>|z|)
Intercept	2.57457	4.57142	0.721	0.471
x1	0.01746	0.05402	0.323	0.747
x2	12.51909	13.09931	0.956	0.339
x3	-2.97388	0.75373	-3.946	7.96e-05
x4	-1.03425	0.66730	-1.550	0.121

Table 7. Binary logistic regression, Fd Cabuya

	Estimate	Std. Error	z value	Pr(>|z|)
Intercept	-1.26567	3.31540	0.385	0.7004
x1	0.10677	0.05315	2.009	0.0446
x2	-18.54705	12.52257	-1.480	0.1389
x3	-1.62258	0.65296	-2.485	0.0130
x4	-0.56086	0.61796	-0.908	0.3641

Likewise, the factors that intervene in the surface quality of the CGFRP are the following (**table 5 and 7**). The cutting speed and orientation are clearly significant when evaluating the Ra, On the other hand, the cutting speed and tool are significant when evaluating the delamination factor. Finally, the following predictive logarithmic equations for (Ra) and (Fd) respectively are obtained:

Greecomposite with Abacá:

$$Z_i = 299(\textit{cutting speed}) - 598(\textit{tool}) \qquad \pi_i = \frac{e^{299(\textit{cutting speed})-598(\textit{tool})}}{1+e^{299(\textit{cutting speed})-598(\textit{tool})}}$$

$$Z_i = -2.97388(\textit{tool}) \qquad \pi_i = \frac{e^{2.97388(\textit{tool})}}{1+e^{2.97388(\textit{tool})}}$$

Greencomposite with Cabuya:

$$Z_i = -0.21427(\textit{cutting speed}) - 257127(\textit{Orientation}) \qquad \pi_i = \frac{e^{-0.21427(\textit{cutting speed})-257127(\textit{Orientation})}}{1+e^{-0.21427(\textit{cutting speed})-257127(\textit{Orientation})}}$$

$$Z_i = 0.10677(\textit{cutting speed}) - 1.62258(\textit{Tool}) \qquad \pi_i = \frac{e^{0.10677(\textit{cutting speed})-1.62258(\textit{Tool})}}{1+e^{0.10677(\textit{cutting speed})-1.62258(\textit{Tool})}}$$

4. **Conclusion**

Experimental analysis shows that machining with the HSC carbide tool gives us a much better surface quality of grooving than the conventional HSS tool, as the primary effects figures show. The binomial logistic regression method was used to determine the appropriate parameters for the grooving of the mechanized biocomposites. In other words, for the AGFRP the use of a cutting speed of 60m/min is obtained, the minimum surface roughness (Ra) is obtained and the use of an HSC tool results in a minimum delamination factor (Fd). Likewise, the surface quality of the composite hybrid CGFRP, is subject to the use of a cutting speed of 48m/min, to obtain a minimum surface roughness (Ra) and a cutting speed of 60m / min to obtain a delamination factor minimum (Fd), and the use of an HSC tool to obtain a minimum delamination factor (Fd).

References

[1] M. P. Lucena, A. Suarez, and I. Zamudio, "Desarrollo de un material compuesto a base de fibras de bambú para aplicaciones aeronáuticas," *Revista Latinoamericana de Metalurgia y Materiales,* pp. 1107-1114, 2009.

[2] V. Guerrero, P. Pontón, A. Tamayo, H. Villacís, F. Delgado, and S. Galeas, "Materiales compuestos de matriz poliéster reforzados con fibras naturales y sintéticas," in *Congreso de Ciencia y Tecnología ESPE,* 2013, pp. 70-74.

[3] J. G. P. Salinas, C. F. P. Salinas, and C. B. C. Miniguano, "Análisis de las propiedades mecánicas del compuesto de matriz poliéster reforzado con fibra de vidrio 375 y cabuya aplicado a la industria automotriz," *Enfoque UTE,* vol. 8, pp. pp. 1-15, 2017.

[4] G. D. Babu, K. S. Babu, and B. U. M. Gowd, "Effect of machining parameters on milled natural fiber-reinforced plastic composites," *J. Adv. Mech. Eng,* vol. 1, pp. 1-12, 2013.

[5] I. El-Sonbaty, U. Khashaba, and T. Machaly, "Factors affecting the machinability of GFR/epoxy composites," *Composite structures,* vol. 63, pp. 329-338, 2004.

[6] J. L. Merino-Pérez, R. Royer, E. Merson, A. Lockwood, S. Ayvar-Soberanis, and M. B. Marshall, "Influence of workpiece constituents and cutting speed on the cutting forces developed in the conventional drilling of CFRP composites," *Composite Structures,* vol. 140, pp. 621-629, 2016/04/15/ 2016.

[7] N. Abilash and M. Sivapragash, "Optimizing the delamination failure in bamboo fiber reinforced polyester composite," *Journal of King Saud University - Engineering Sciences,* vol. 28, pp. 92-102, 2016/01/01/ 2016.

[8] A. P. Singh, M. Sharma, and I. Singh, "A review of modeling and control during drilling of fiber reinforced plastic composites," *Composites Part B: Engineering,* vol. 47, pp. 118-125, 2013/04/01/ 2013.

[9] A. Turon, J. Costa, P. Camanho, and C. Dávila, "Simulation of delamination in composites under high-cycle fatigue," *Composites Part A: applied science and manufacturing,* vol. 38, pp. 2270-2282, 2007.

[10] M.-C. Francisco, H. Issam, K. Abdellatif, J. Abdallah, and B. Mohamed, "Predicción de rugosidad en maquinado de compuestos con base de Peek usando metodología de superficie de respuesta," *Ingeniería, investigación y tecnología,* vol. 14, pp. 463-474, 2013.

[11] K. Palanikumar, L. Karunamoorthy, and R. Karthikeyan, "Parametric optimization to minimise the surface roughness on the machining of GFRP composites," *Journal of Materials Science and Technology,* vol. 22, pp. 66-72, 2006.

[12] H. S. S. Aljibori, W. Chong, T. Mahlia, W. Chong, P. Edi, H. Al-Qrimli, *et al.*, "Load–displacement behavior of glass fiber/epoxy composite plates with circular cut-outs subjected to compressive load," *Materials & design,* vol. 31, pp. 466-474, 2010.

[13] R. Kishore, R. Tiwari, A. Dvivedi, and I. Singh, "Taguchi analysis of the residual tensile strength after drilling in glass fiber reinforced epoxy composites," *Materials & design,* vol. 30, pp. 2186-2190, 2009.

[14] P. Benardos and G.-C. Vosniakos, "Predicting surface roughness in machining: a review," *International journal of machine tools and manufacture,* vol. 43, pp. 833-844, 2003.

[15] R. Vinayagamoorthy and N. Rajeswari, "Analysis of cutting forces during milling of natural fibered composites using fuzzy logic," *Int J Compos Mater Manuf,* vol. 2, pp. 15-21, 2012.

[16] N. Muthukrishnan and J. P. Davim, "Optimization of machining parameters of Al/SiC-MMC with ANOVA and ANN analysis," *Journal of materials processing technology,* vol. 209, pp. 225-232, 2009.

[17] D. Niranjan, G. Shivashankar, K. S. Rao, and R. Praveen, "Optimization of Cutting Process Parameters on AL6061 Using ANOVA and TAGUCHI Method," *Materials Today: Proceedings,* vol. 4, pp. 10845-10849, 2017.

[18] P. Palanisamy, I. Rajendran, and S. Shanmugasundaram, "Prediction of tool wear using regression and ANN models in end-milling operation," *The International Journal of Advanced Manufacturing Technology,* vol. 37, pp. 29-41, April 01 2008.

[19] J. Lee, L. Liao, E. Lapira, J. Ni, and L. Li, "Informatics Platform for Designing and Deploying e-Manufacturing Systems," in *Collaborative Design and Planning for Digital Manufacturing*, L. Wang and A. Y. C. Nee, Eds., ed London: Springer London, 2009, pp. 1-35.

[20] J. Karandikar and T. Kurfess, "Cost optimization and experimental design in milling using surrogate models and value of information," *Journal of Manufacturing Systems,* vol. 37, pp. 479-486, 2015/10/01/ 2015.

[21] S. Sperandei, "Understanding logistic regression analysis," *Biochemia Medica,* vol. 24, pp. 12-18, 02/15/09/06/received11/26/accepted 2014.

[22] G. Nassimbeni, "Technology, innovation capacity, and the export attitude of small manufacturing firms: a logit/tobit model," *Research Policy,* vol. 30, pp. 245-262, 2001/02/01/ 2001.

[23] A. Carpinteri, A. Fernández-Canteli, G. Fortese, M. Muñiz-Calvente, C. Ronchei, D. Scorza, *et al.*, "Probabilistic failure assessment of Fibreglass composites," *Composite Structures,* vol. 160, pp. 1163-1170, 2017/01/15/ 2017.

[24] E. Glistau and N. I. C. Machado, "Design Of Experiments And Six Sigma Methods In Logistics," *Advanced Logistic systems,* vol. 2, pp. 47-54, 2008.

[25] H. G. Villacís Salazar, "Obtención de materiales compuestos híbridos de matriz poliéster reforzados con fibra de vidrio y abacá mediante estratificación," QUITO/EPN/2011, 2011.

IOP Conf. Series: Materials Science and Engineering **446** (2018) 012004 doi:10.1088/1757-899X/446/1/012004

A Novel Nitrogen-Assisted Method for the Perovskite Active Layer Optimization in the Inverted Planar Heterojunction Perovskite Solar Cells

Miao Yu [1]*, **Xiaona Huang** [2], **Junze Shao** [1], **Wanli Zhang**, [1]

[1]State Key Laboratory of Electronic Thin Films and Integrated Devices, University of Electronic Science and Technology of China, Chengdu, China
[2]Chengdu Technological University, Chengdu, China

* E-mail: yumiao@uestc.edu.cn

Abstract. The perovskite solar cells is recognized as a promising candidate for energy industry system because of its tunable band-gap, high absorption coefficient, broad absorption spectrum, high charge carrier mobility and long charge diffusion lengths. However, a dendritic grains morphology, caused by the spin coating procedure, will result in poor coverage and then low power conversion efficiency. In this work, a novel and facile inactive gas assisted method is introduced to obtain uniform and dense perovskite thin films. The dry nitrogen gas blows over the surface of the perovskite solution during the spin-coating process, which promote a faster solvent evaporation. Then, the uniform and compact perovskite film on the substrate was achieved. Typically, the power conversion efficiency (PCE) of the nitrogen gas assisted perovskite solar cells increases from 0.30% to around 9.5 %. Furthermore, the favorable start time of nitrogen blowing on devices was also investigated. The PCE stays around 9.46 % – 9.78 % when the gas starts at 2 sec - 5 sec, and the efficiency decreases to 8.98 % if the gas starts at 6 sec. The nitrogen flow assisted method is proved to be an effective way to fabricate dense perovskite films without polluting the environment.

1. Introduction

Organic-inorganic hybrid perovskite materials, applied in the perovskite solar cells, have shown great promise in field-effect transistors, light-emitting diodes, sensors, and photodetectors [1] due to their tunable band-gap, high absorption coefficient, broad absorption spectrum, high charge carrier mobility and long charge diffusion lengths [2]. Recently, the power conversion efficiency (PCE) of lead halide perovskite (CH3NH3PbX3, X = Cl, Br, I) based thin film photovoltaic devices has boost from 3.8% [3] to more than 22.1% [4], [5] in just eight years. The PCE is highly determined by the light-absorbing film's deposition process which is strongly connected with the morphology and coverage ratio. Usually, perovskite films used in solar cells have been fabricated by a co-evaporation method [6], two-step [7], [8] or one-step [7], [9] solution process. The vapor deposition process is likely to increase the manufacturing cost whereas the sequential two-step deposition procedure will involve a longer overall processing time [10], [11]. The one-step solution deposition is a fast and simple fabrication method but always causes low surface coverage. Xiao et al. developed a fast crystallization-deposition (FDC)

IOP Conf. Series: Materials Science and Engineering **446** (2018) 012004 doi:10.1088/1757-899X/446/1/012004

method that involves the spin-coating of a N,N-dimethylformamide (DMF) solution of CH3NH3PbI3 on a substrate, followed immediately by the exposure of the wet film to a second chlorobenzene (CB) solvent to fabricate high quality planar CH3NH3PbI3 thin films, such as full surface coverage, small surface roughness, microscale grain size, and high PCE of 16.2% [10]. Han et al. reported a dissolution-recrystallization method (DRM) with chlorobenzene and dimethylsulfoxide treating the perovskite films during the spin-coating process to control perovskite crystallization and improve the device performance [12]. However, many solvents are poisonous and environmental-unfriendly. Therefore, it is a necessity to develop an alternate method.

Herein, we reported a novel inactive gas (nitrogen gas) assisted technology, which offers the advantage of single-step processing and environmental-friendship. This simple technology involves the spin-coating of a DMF solution of CH3NH3PbI3 on a substrate, followed immediately by exposure of the wet film to a constant nitrogen flow to induce high crystalline single grains. By this new technology, planarheterojunction solar cells yielded a maximum PCE of 9.78% under standard AM 1.5 conditions. We believe this facile method can be applied for the fast production of highly efficient CH3NH3PbI3 based solar cells.

2. Experimental

2.1 Experimental materials

Poly 3,4-ethylenedioxythiophene: poly styrenesulfonate (PEDOT:PSS) and Clevios^{TM} PVP. AI 4083 are purchased from Heraeus Materials Technology Shanghai Ltd. CH3NH3I (MAI) was synthesized via the approach described in the literature [8]. Lead iodide (PbI2, 99%) and Anhydrous N,N-dimethylformamide (DMF, 99.8%) are purchased from Sigma-Aldrich company. [6,6]-phenyl-C61-butyric acid methyl ester (PC61BM,>99%) was purchased from Xi'an Polymer Light Technology Corp. To prepare the perovskite precursor solution, CH3NH3I and PbI2 were mixed in DMF at the molar ratio of 1:1, respectively. The solution was stirred at 60°C overnight, and was then filtered using 0.45 μm PTFE filters before being used for device fabrication. The concentration of the perovskite precursor solution was 40 wt%.

2.2 Device fabrication

The device structure is shown in Figure 1, starting from the bottom: Glass/ITO substrate, hole transport layer PEDOT:PSS, perovskite photoactive layer, and PC61BM electron transport layer, with a final Ag electrode deposited via thermal evaporation.

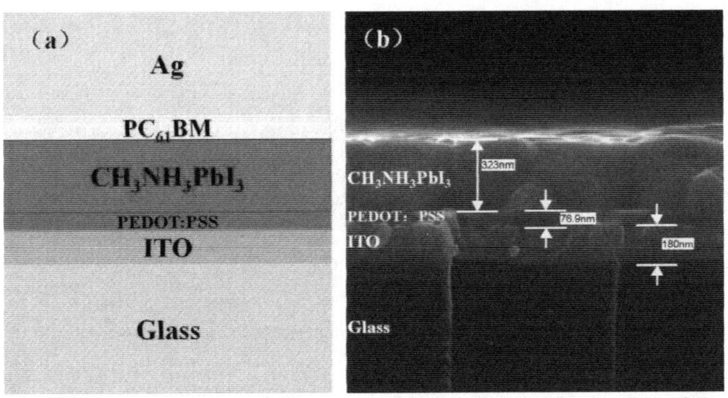

Figure 1. Schematic illustration (a) and Cross-sectional SEM image (b) of the designed device.

The device is fabricated according to the previous report [13], [14]. Indium tin oxide (ITO, 15 ohm/sq) glass substrates were cleaned sequentially under ultrasonication for 20 min with detergent, deionized water and isopropyl alcohol. Then PEDOT: PSS films were deposited on the ITO coated glass by spin-coating the PEDOT: PSS solution at a speed of 5000 rpm. After heated at 150 °C on a hot plate for 10 min, the PEDOT:PSS films were mild argon plasma post-treated with PDC-36 G under a base pressure of around 170 Pa for 3 min. The CH3NH3PbI3 perovskite active layer was deposited onto the as-prepared PEDOT: PSS layer by spin-coating CH3NH3PbI3 precursor solution (40 wt% in DMF) at 4000 rpm for 40 s. In order to improve the crystallization of perovskite active layer deposited by one-step solution method, an additional step was introduced where a 0.4Mpa dry nitrogen gas flow was blown over the surface of the perovskite solution during the spin-coating process. Then the films were annealed at 110 °C for 30 min inside the glove box filled with nitrogen. After that, a solution of PC61BM in CBZ was spin-coated onto the perovskite films at 3000 rpm for 40 s. Finally, a silver electrode (100 nm) was deposited by thermal evaporation through a shadow mask under a base pressure of 3×10-4Pa, which defined an active cell area of 0.075cm2.

2.3 Characterization

To improve the measurement accuracy, the J–V curves for all devices were measured by masking the active area with a metal mask. The device photocurrent was measured under an AM 1.5G solar simulator (Japan, SAN-EI, XES-40S1). The current density-voltage (J-V) characteristics for the devices were recorded with a Keithley 2400 source meter and no poling or light soaking process was applied before the J-V measurement. The illumination intensity of the light source was calibrated before the tests by using a standard silicon solar cell with a KG5filter, which was calibrated by a National Renewable Energy Laboratory (NREL) silicon photo diode, giving a value of 100 mW cm-2. All J-V curves reported in this study were measured under a reversed voltage bias. Average values and standard deviations were generated from the measurements of these spots. An X-ray diffractometer (XRD, Philips 1140) with Cu Ka radiation was utilized to compare the crystalline structures of various devices with and without nitrogen assisted. The surface and cross section SEM images were recorded with a FEI Nova dual beam, focused ion beam (FIB), SEM instrument using 5 kV acceleration voltage. Prior to ion beam thinning of the cross section, two platinum protecting layers were deposited in situ using an electron beam source at 6.3 nA and an ion beam source at 0.30 nA. The milling of the cross sections was performed with a gallium ion source at a 52-tilting angle.

3. Results and Discussion

3.1 Procedure of the gas-assisted method

The process of gas assisted method for preparing the CH3NH3PbI3 films is shown in Figure 2. A DMF solution of CH3NH3PbI3 (40wt%) was spin-coated on the PEDOT:PSS layer at 4000 rpm. Compared to the conventional method, a fast nitrogen flow with 0.4 M pa pressure quickly blew vertically to the substrate.

The role of fast nitrogen flow is to reduce the solubility of CH3NH3PbI3 in the pristine solvent rapidly and thereby promote fast nucleation and growth of the crystals in the film. In contrast, during a conventional spin-coating process where no fast nitrogen flow was involved, the wet film dried slowly and a whitening-gray film was produced. The films were then annealed at 120 °C for 30 min to evaporate any residual solvent and promote crystallization. As shown in Figure 2, the nitrogen gas processed films become darker which indicates the formation of the compact thin film.

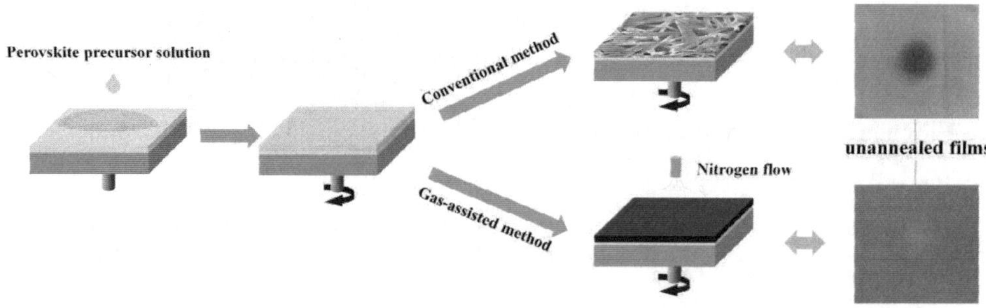

Figure 2. Schematic procedure for the gas-assisted method.

3.2. Effect of gas assisted method on perovskite films

The nitrogen flow that blows over the surface of the CH3NH3PbI3 wet film can promote the solvent evaporation and accelerate the supersaturation of the CH3NH3PbI3 precursor solution. The influence of start time on film properties (the beginning of spin-coating defined as starting time of 0 sec) were tested. Figure 3 shows the SEM images of surface morphology of CH3NH3PbI3 films treated by the nitrogen flow at different start time. It is found that the film prepared by conventional method resulted in a dendritic structure that poorly covers the underlying surface (Figure 3.a), while the nitrogen assisted film shows a smoother surface (Figure 3b-e). As the start time varied from 2 to 5 sec, film surface shows dense micro-structure. When the start time increased to 6 sec, the dendritic grain morphology were produced.

In general, the removal of excess precursor solution is a dominant process in the first 2 sec after spin-coating was commenced [15]. The solvent of DMF has high boiling point(153 °C) and low vapor pressure (2.6 mmHg at 20 °C). For the conventional method, slow evaporation results in a low degree of super-saturation, which favors the crystal growth of a small number of existing nuclei, thereby promoting the formation of dendritic perovskite grains [16] (Figure 3a). Whereas, nitrogen flow causes a high degree of super-saturation, the solution is saturated enough to form homogeneous crystallization [17], which can enhance crystallization and prefer-orientation of perovskite crystal (Figure 3b-e). When the time reached 6 sec, the start time is too late that a small number of dendritic grains in random orientation have already been produced (Figure 3f).

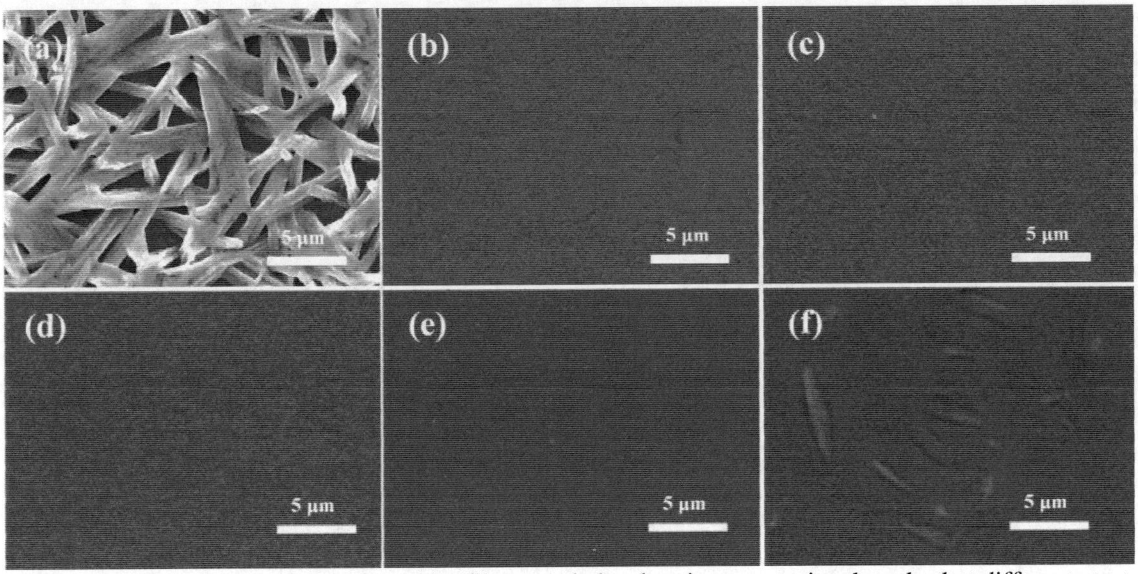

Figure 3. SEM images of the perovskite films made by the nitrogen-assisted method at different start time. (a) untreated, (b) 2 sec, (c) 3 sec, (d) 4 sec, (e) 5 sec, (f) 6 sec.

XRD analysis was used to further investigate the influence of blowing start time on the film composition. As shown in Figure 4, there was no obvious difference between 2 and 4 sec, but a clear reduction in the intensity of the CH3NH3PbI3 perovskite diffraction peaks (14.1°, 28.4, 32°) as the start time reaches 6 sec. That indicated a possible change of the grain crystal orientation and granulation of perovskite film, which is consistent with the SEM results.

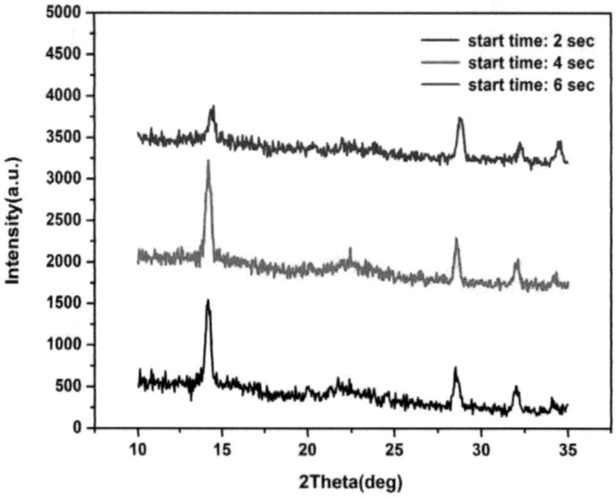

Figure 4. XRD pattern of perovskite films made by the nitrogen-assisted method at different start time.

3.3. The Performance of the devices

Figure 5 shows the photovoltaic parameters of devices fabricated with the nitrogen assisted method. Different start time results in devices with different PCE parameters. The cells show PCEs improvement after nitrogen blowing. However, as the start time is later than 5 secs, the PCEs show a decreasing trend. Uneven coating of perovskite (Figure 3f) causes an uneven coating of the overlaying of PCBM layer. The non-uniform perovskite films exhibited regions with extremely thin coverage of the substrate. In these regions the shunt resistance would be reduced resulting in a lower voltage and fill factor [16]. Therefore, the best start time to produce device with high PCE should be between 2 - 5 sec.

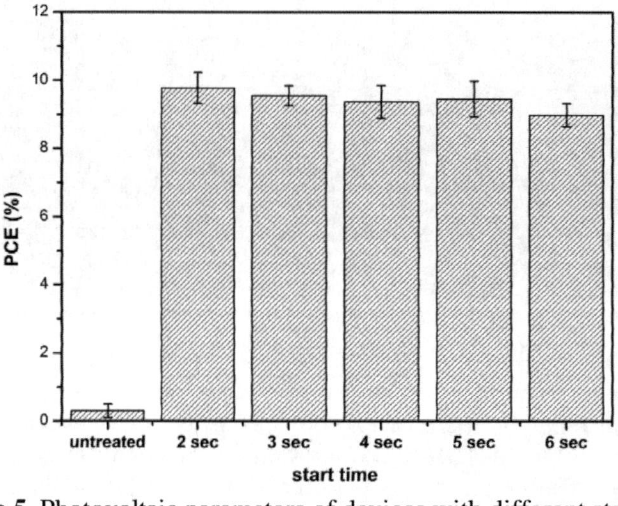

Figure 5. Photovoltaic parameters of devices with different start time.

4. Conclusion

In summary, nitrogen assisted method has been developed to fabricate high quality of perovskite films. Compared to the conventional process, this new technique leads to a higher degree of super-saturation resulting the nucleation more favorable that make a uniform perovskite film fully covers the substrate. Furthermore, the gas flow blowing start time from the beginning of spin-coating has also been investigated. When the start time varies between 2 and 5 sec, there are no dendritic grains. The PCE of perovskite solar cells increased from 0.30% for the pristine CH3NH3PbI3 films based device to 9.78%,9.55%,9.37% and 9.46% by the nitrogen-assisted method delayed for 2 sec, 3 sec, 4 sec, 5 sec, respectively. Gas-assisted method is proved to be a useful plan to fabricate dense perovskite film without harm influence to environment.

Reference

[1] Zhou H, Chen Q, Li G, et al. Photovoltaics. Interface engineering of highly efficient perovskite solar cells.[J]. Science, 2014, 345(6196):542-6.
[2] Pierre A, Arias A C. Solution-processed image sensors on flexible substrates[J].2016, 1(4).
[3] A. Kojima, K. Teshima, Y. Shirai and T. Miyasaka, J. Am.Chem. Soc., 2009, 131, 6050–6051.
[4] http://www.nrel.gov/ncpv/images/efficiency_chart.jpg
[5] Yang W S, Park B W, Jung E H, et al. Iodide management in formamidinium-lead-halide-based perovskite layers for efficient solar cells[J]. Science, 2017, 356(6345):1376-1379.
[6] Liu M, Johnston M B, Snaith H J. Efficient planar heterojunction perovskite solar cells by vapour deposition.[J]. Nature, 2013, 501(7467):395-8.
[7] Im J H, Kim H S, Park N G. Morphology-photovoltaic property correlation in perovskite solar cells: One-step versus two-step deposition of CH3NH3PbI3[J]. Apl Materials, 2014, 2(8):081510.
[8] Chen Q, Zhou H, Hong Z, et al. Planar Heterojunction Perovskite Solar Cells via Vapor-Assisted Solution Process[J]. Journal of the American Chemical Society, 2014, 136(2):622.
[9] Rhee J H, Chung C C, Diau W G. A perspective of mesoscopic solar cells based on metal chalcogenide quantum dots and organometal-halide perovskites[J]. Npg Asia Materials, 2013, 5(10):e68.
[10] Xiao M, Huang F, Huang W, et al. A fast deposition-crystallization procedure for highly efficient lead iodide perovskite thin-film solar cells[J]. AngewandteChemie, 2014, 53(37):9898.
[11] J. Burschka, N. Pellet, S.-J. Moon, R. Humphry-Baker, P. Gao,M. K. Nazeeruddin, M. Gr tzel, Nature 2013, 499 , 316.
[12] Han F, Luo J, Wan Z, et al. Dissolution-recrystallization method for high efficiency perovskite solar cells[J]. Applied Surface Science.
[13] You, J. et al. Low-temperature solution-processed perovskite solar cells with high efficiency and flexibility. ACS Nano 8, 1674–1680 (2014).
[14] Docampo, P., Ball, J. M., Darwich, M., Eperon, G. E. &Snaith, H. J. Efficient organometaltrihalide perovskite lanar-heterojunction solar cells on flexiblepolymer substrates. Nature Commun. 4, 2761 (2014).
[15] Snaith H J, Humphry-Baker R, Chen P, et al. Charge collection and pore filling in solid-state dye-sensitized solar cells[J]. Nanotechnology, 2008, 19(42):424003.
[16] Huang F, Dkhissi Y, Huang W, et al. Gas-assisted preparation of lead iodide perovskite films consisting of a monolayer of single crystalline grains for high efficiency planar solar cells[J]. Nano Energy, 2014, 10:10-18.
[17] Baikie T. Synthesis and crystal chemistry of the hybrid perovskite (CH3NH3)PbI3 for solid-state sensitised solar cell applications[J]. Journal of Materials Chemistry A, 2013, 1(18):5628-5641.

IOP Conf. Series: Materials Science and Engineering **446** (2018) 012005 doi:10.1088/1757-899X/446/1/012005

Development of Maintenance-Reparation Guidance, and Material Specification for Green Building's Electrical Component in Government Building Based on Work Breakdown Structure (WBS)

Resha Mochamad Ilham, Yusuf Latief, Leni Sagita Riantini, Budi Susilo

Department of Civil Engineering, Indonesia University, Jakarta, Indonesia

Email: resha.moch.ilham@gmail.id, latie73@eng.ui.ac.id, lsagita@eng.ui.ac.id, budisus@eng.ui.ac.id

Abstract. Initial quality of building depands on quality of design and construction phase of the building, while maintenance and spare part quality keep the decline of the quality during operational phase of building within level of service needed. Electrical aspect of building has an impact of the building's purpose, safety, and comfort. Several finding in operation phase have been found regarding lack of quality of electrical maintenance process and spare part material such as fire incident, waste of electricity and exesive stock of spare part. Thus, it's important to not only develop maintenance and reparation guidelines of electrical work but also it's material specification to keep the required level of service. The purpose of this study is to use work breakdown structure (WBS) methods as a proven method in decomposition to develop maintenance and reparation guidelines, as well as the material specifications in green building. The research methodology covers several stages: literature studies, information from previous projects, and delphi techniques through expert validation in green building electrical components. The results of this study are expected to create applicable guidelines for maintenance and reparation work of electrical component and it's material specification to eliminate or decrease finding of electrical issues in building operational.

1. Introduction

According to Indonesia Centre Bereau of Statistic, the estimated value of construction works of the completed buildings is more than Rp. 206 Trillion in 2017. The value contributes almost 30% of the total construction industry in Indonesia (Rp 696T) and 1.7% of the total value of Gross Domestic Product (Rp 12.406T). The vast amount of the value demands high quality and level of service to the occupant.

Development of the building shall comply with government regulation such as Regulation of Public Works dan Housing Minister (PUPR) No 02/PRT/M/2015 and Jakarta Governor Regulation No 38/2012 that expressed its importance of green building method in design, construction and operational phase of the building. In addition to the regulation, green building method not only gives benefit to occupant's productivity and health [1], comfort and statisfaction [2], but also to the building owner and management in the form of lower operational and energy cost and effective energy utilization [3].

Aplication of green building by various independent certifications are based on criteria such as Site, Energy, Water, Indoor Enviromental Quality (IEQ), Material, Waste & Polution, and Management, with the most influential criteria on the assessment is the energy aspect[4]. Electrical as the most common energy use in building should have biggest impact on Green Building application. Miss management of the electrical could leads to building incident such as in efficient use of electricity [5], short circuit or fire [6], and spare part overstock [7].

These electrical impacts of building raise the needs of proper maintenance and material specification to make sure quality of the building. Designers / Engineers play an important role to produce designs with good quality and construcability [8]. The maintenance guideline and material specification reduce dependence on personal in order to keep quality of the building as per required in every phase of maintenance of building.

The scope of this study are limited to the following : (1) standardized WBS for electrical aspect in the building, (2) standardized electrical material specification, and (3) the research conducted on green building project in government building.

2. Research objectives

The objectives of this research are :

1. To identify the type and work package for maintenance and reparation in government green building
2. To identify alternative design, activity, implementation requirement, and resources for government green building maintenance and reparation
3. To determine material specification
4. To develop guideline for maintenance and reparation of electrical aspect

3. Literature review

3.1. Government buildings

According to Minister of Public Works Regulation no. 45 Year 2007, Goverment Building is a building for the purposes of service that becomes / will become state-owned property such as office buildings, school buildings, hospital buildings, warehouses, and country houses, and held with financing sources derived from state budget funds, and / or other legitimate acquisitions.

3.2. Green building

According to Regulation of Public Works and Housing Minister (PUPR) No.2 year 2015 on green building, green building is a building that fulfills the building requirement and has significantly measured performance in energy, water and other resource savings through application principles of green building according to function and classification in each stage of its implementation. There are national and international organizations that have own definition and criteria to certify a building as green building.

3.3. Work break down structure

Work Breakdown Structure (WBS) as a proven method to decomposed higher component to the smallest component is used to create the structured and completed guidelines. WBS is a structured decomposition of the overall scope of work that the project team must accomplish to accomplish the project objectives and complete the required results [9]. A complete and efficient WBS will be an important part of project management planning by dividing the project into multiple stages, deliverables and work packages [10]. There are several efforts to standardized WBS such as underground construction [10], apartment building [11], high rise building architectural works [12] and building safety plan [13]. This study will employ WBS method in maintenance and reparation works in government green building..

Proposed WBS stage / component in electrical maintenance and reparation works is shown in Figure 1.

3.4. Maintenance and reparation guidelines

Final goals of this study is applicable maintenance and reparation guidelines, which consist of alternative design in all work package, the activity that will be implemented in each equipment/alternative design, procedure and resources to complete the activity and material spesification for spare part or replacement unit.

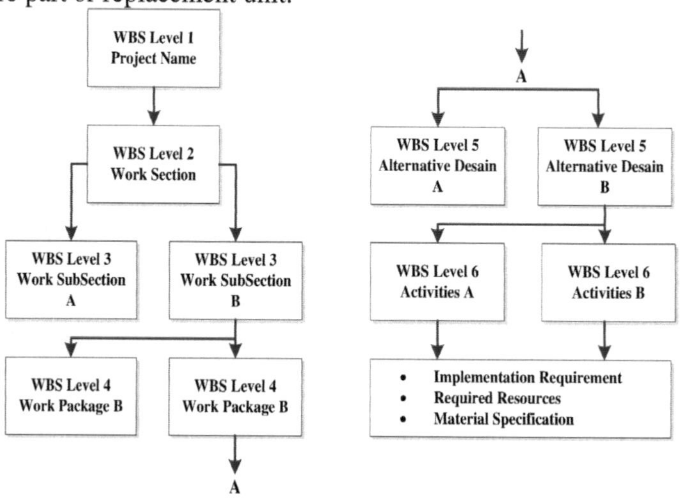

Figure 1. WBS diagram tree.

4. Research methodology

Qualitative research method is used in this study to get result in each reseach objective. (1) central and local government regulation, project archieve, green building certification program reference and other journal or books are used to identify sub work section and work package, the the data are validated by expert using delphi method. (2) Further study literature is used to determine draft of alternative design, activity, implementation requirement, resources needed to completed maintenance and reparation of each work package, that draft shall be validated by the expert. (3) furthermore, material specification determined by literature studies, with government regulation, national standard, green building certification requirement, and past project experience take into consideration. Validation by expert conducted to compile the final material specification (4) previous result of research objective collected and docummented as maintenance and reparation guideline, then final validation through experts use delphi method.

5. Results and discussion

WBS in figure 2 has been validated by expert, decommposed all required work package in maintenance and reparation of electrical work in government green bulding. One of differientiation between typical office building with green office building is renewable energy generation requirement in green building, this item can be found in the sub work section and work package.

The WBS devided by four sub work section, and twenty work packages. First sub work section is power sources as the system that generates power, or as receiver from grid system, consists of transformator, UPS, backup power generator, and renewable energy power generation work package. Second sub work section is distribution, as the system that transmit electrical form source to load, consist of medium voltage panel and LVMDB Panel work package.

Third sub work section is electrical load, consists electrical load panel, cable, lighting equipment, lighting control system, receptacle and switch, grounding system and lightning protection system work packages. Last sub work section is electronics consists of fire alarm and detectors, telephone, sound system, internet and computer, TV and CCTV, signal repeater and building automation system work packages.

Components of the building will be differing for each building depends of the design of the building. This study will present alternative design available for each package. One of the workpackage is shown below,

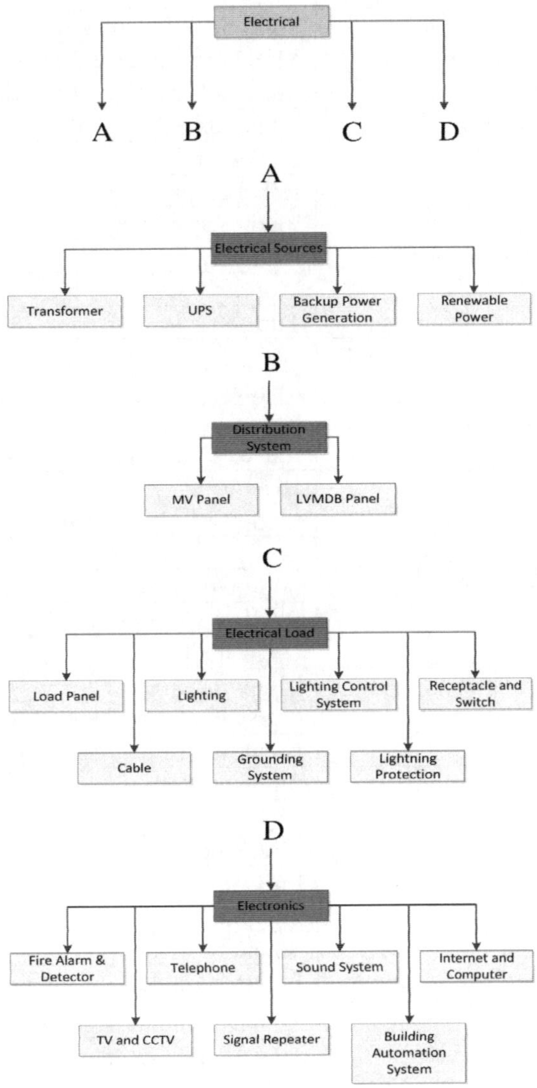

Figure 2. Electrical sub work section and workpackages.

Table 1. Alternative design of electrical works.

Work Package WBS Level 3	Alternatif Design WBS Level 4
Transformer	Dry Transformer
	Liquid Transformer
Uniterruptable Power Supply	Local UPS
	Central UPS
Backup Power Sources	Diesel Engine Generator
	Gas Engine Generator
Renewable Energy Sources	Solar Panel
	Ground Sources Heat Pump
	Wind Turbine
Medium Voltage Panel	Oil Circuit Breaker
	Vaccum Circuit Breaker
Low Voltage Main	Mechanical Circuit Breaker

Distribution Panel	Solid State Circuit Breaker
	Miniature Circuit Breaker
	Molded Case Circuit Breaker
Load Panel	Load Panel
Artificial Lighting	LED Lamp
	HID Lamp
	Fluorescent Lamp
Ligthting Control System	Dimmer
	Ligthting Sensor
	Time Sensor
Receptacle and Switch	Receptacle
	Switch
Under Floor Duct / Raised Floor	Under Floor Duct
	Raised Floor System
Cable	High Voltage Cable
	Medium Voltage Cable
	Low Voltage Cable
Grounding System	Solid Grounding
	Resistor Grounding
	Ungrounding
Lightning Protection	Lightning Rod
	Lightning Rod - Tight Wire
	Lightning Rod - Meshed Cage
Fire System and Detector	Smoke Detector
	Heat Detector
	Fire Panel
	Fire Alarm
Telephone	Telephone
Sound System	Sound System
Internet and Computer System	Ethernet Cable
	Wifi
TV dan CCTV	TV
	CCTV
Signal Repeater	Signal Repeater GSM
Building Automation System	Sensor
	Control Panel

This list of alternative design covered all posibility of work packags needed in electrical works are shown in Tabel 1. Each of alternative design will need a specific implementation requirement formed from action to be taken from each activity and resources to complete the activity.

Table 2 shows example of one alternative design including it's detail of activity and resources.

Table 2. Activity and resources of solar panel work packages

WBS Level 1 Maintenance & Reparation Works WBS Level 2 Electrical WBS Level 3 Power Source WBS Level 4 Renewable Energy Sources WBS Level 5 Solar Panel	
Inspection Activity	
-Voltage measurement in each solar panel, batere bank, rectifier	Resources: -Electrical Tecnician

-Visual inspection in present of dust, rust, and inclination of solar panel -Check tightness of bolt/nut -Check leak or damage in batere cell	-Multimeter -Torquemeter -Hand Tools
Maintenance Activity	
-Make sure area n devise is safe, if required use LOTO device -Clean solar cell from dust and rust -Fix inclination and solar panel position -Fix torque of bolt and nut -Check level of batere elektrolit	Resources: -Electrical Tecnician -Multimeter -Torquemeter -Hand Tools
Reparation Activity	
-Replace solar panel and panel that exceed lifetime or drop in efficiency -Replace electrical component that broken or obsolete -Fill up the batere electrolite	Resources: -Electrical Tecnician -LOTO -Multimeter -Torquemeter -Hand Tools -Batere Electrolite -Spare Part
Testing Activity	
-Test output voltage of the solar panel -Test grounding condition -Test cable insulation condition -Test electrical protection -Test batere system backup time	Resources: -Electrical Tecnician -LOTO -Multimeter -Insulation Tester -Hand Tools

Then, material specification for each altenative design shall be developed and validated by expert. Tabel 3 shows example of material specification proposed for solar panel

Table 3. Example of material specification

Solar Panel Material Specification
-Comply with SNI 04 3850.2 Tahun 1995 -Comply with latest Panduan Umum Instalasi Listrik – Government Regulation of General Guidance of Electrical Installation -Comply with ISO 14001 -Produce within 1000 KM radius from building site -Equiped with Power Management system software

All this material specification shall be valid for spare part and or unit replacement. Next result of this study is compilation of all previous objectives, Maintenance and Reparation Guidelines. Table 4 shows example of inspection part of maintenance and reparation guidelines of solar panel in green building. This guidelines validated by building owner and operator.

Table 4. Maintenance and reparation of solar panel

RENEWABLE ENERGY RESOURCES
Description : The document covers guideline for maintenance and reparation of renewable energy work packages including its alternative design which is 1.Solar Panel system;

2.Ground Sources Heat Pump system; 3.Wind Turbin system. The general requirement of the overall activity including -Use standard and special personel protective equipment; -Obtain permit and clearence; -Clear and secure area; -Risk assessment and mittigation	
SOLAR PANEL	
Section	Electrical Work
Sub-Section	Power Source
Work Packages	Renewable Power Source
Alternative Design	Solar Panel System
GUIDELINES	
Inspection	
This activities cover surveillence and monitoring of the device or system to find anomaly or defect during normal operation, without shutdown the system or device. The anomaly and defect shall be register and planned for next maintenace or reparation activity.	
Detail Procedure	**Resources**
1.Prepare documentation, permit, and drawing for the system 2.Prepare tools, personal safety equipment and access 3.Prepare baricade and signage 4.Check visual of each of system (Solar Panel, Cabling, Rectifier, Batere System and Distribution Panel) for dust, rust, defect and anomally. 5.Check visual inclination of solar panel module 6.Check voltage and ampere of input-output of each module 7.Check torque of bolt (if posible) 8.Check Anomaly in batere system, if there are any leak, rust, or cracking	1.Engineering Drawing 2.Electrical Technician 3.Tools, Personal Safety Equipment 4.Barricade and Signage 5.Inclination Tools 6.Multimeter 7.Torque Meter
Material Specification	
1.Comply with SNI 04 3850.2 Tahun 1995 2.Comply with latest Panduan Umum Instalasi Listrik – Government Regulation of General Guidance of Electrical Installation 3.Comply with ISO 14001 4.Produce within 1000 KM radius from building site 5.Equiped with Power Management system software	

6. Conclusion

This compiled guideline of maintenance and reparation for government green building should help building owner or operator to maintaine quality and level of service that tenant or occupant needed. There are others aspect of maintenance and reparation in building need such as architecture, interior, landscape, and mechanical that should have maintenance guidenence also. Further study should be consider to create maintenance standarized for non office building such as apartement and specialized building (airport, port, school library etc).

References

[1] Chau, C. K., Tse, M. S., & Chung, K. Y. (2010). A Choice Experiment to Estimate the Effect of Green Experience on Preferences and Willingness-to-pay for Green Building Attributes. Building and Environment, 45(11), 2553–2561.

[2] https://doi.org/10.1016/j.buildenv.2010.05.017
[3] Paul, W.L. dan Taylor, P.A. (2008). A Comparison of Occupant Comfort and Satisfaction Between a Green Building and Conventional Building. Building and Environment. Vol. 43 No. 11 pp 1858-70.
[4] Ying Liu, J., Pheng Low, S., & He, X. (2012). Green Practices in the Chinese Building Industry: Drivers and Impediments. *Journal of Technology Management in China*, *7*(1), 50–63. https://doi.org/10.1108/17468771211207349
[5] Illankoon, I.M.C.S., et all., (2017). *Key Credit Criteria Among International Green Building* Rating *Tools,* Journal of Cleaner Production Vol.164 pp. 209-220
[6] Agustinus, M. (2016). Pemborosan Listrik di Gedung-gedung Pemerintah Capai 30%. Detik. Retrieved from https://finance.detik.com/energi/3176383/pemborosan-listrik-di-gedung-gedung-pemerintah-capai-30
[7] Nugroho, S. A. (2018). Kebakaran Museum Bahari Diduga akibat Korsleting Listrik.
[8] Wang, W. (2012), " A stochastic model for joint spare parts inventory and planned maintenance optimisation", *European Journal of Operational Research,* Vol. 316 No.1, pp. 127-139.
[9] Au-yong, C. P., Ali, A. S., & Ahmad, F. (2016). Enhancing Building Maintenance Cost Performance with Proper Management of Spare Parts. Journal of Quality in Maintenance Engineering, 22(1), 51–61. https://doi.org/10.1108/JQME-01-2015-0001
[10] Project Managemetn Institute, (2018). "Project *Management Body of Knowledge"* Third Edition
[11] Siami-Irdemoosa, E., et all (2015). Work Breakdown Structure (WBS) development for Underground Construction. Jurnal of Automation in Construction Vol. 58 pp 85-94
[12] L. Sagita, Y. Latief, B. Susilo, M. Rajasa, "*Development of Risk-Based Standardized WBS (Work* Breakdown *Structure) for Cost Estimation of Apartment Project".* International Journal of Civil Engineering and Technology. Volume 8 pp 822-833
[13] Rianty, M., Latief Y., Riantini, L.S., "Development of Risk-based Standardized WBS (Work Breakdown Stuctured) for Quality Planning of High Rise Building Architectural Works".
[14] Elsye, V., Latief, Y., Sagita, L., " Development of Work Breakdown Structure (WBS) Standard for Producing Risk Based Structural Work Safety Plan of Building"

IOP Conf. Series: Materials Science and Engineering **446** (2018) 012006 doi:10.1088/1757-899X/446/1/012006

Cold drawn pearlitic steel wires for wind turbines structures: *In the wake of Miguel de Cervantes and Johann Sebastian Bach*

Prof. Dr. Eng. Jesús Toribio

Fracture & Structural Integrity Research Group (FSIRG)
University of Salamanca (USAL); EPSZ, Campus Viriato,
Avda Requejo 33, 49022 Zamora, Spain

E-mail: toribio@usal.es

Abstract. Wind turbines are key elements in the field of renewable energies. They can be considered as engineering evolutions of old-fashion windmills appearing in the master work by Miguel de Cervantes: *Don Quijote de la Mancha*. The main subject of this paper is related to fracture behaviour and structural integrity of cold drawn pearlitic steel wires for wind turbine structures and foundations. At the microscopic level, cold drawing generates progressive slenderizing and orientation of the pearlitic colonies (*first microstructural level*) and increasing orientation and densification of the ferrite/cementite lamellae (*second microstructural level*), thereby inducing anisotropic fracture behaviour and crack path deflection. In addition, the hierarchical structure of cold drawn pearlitic steel wires in two microstructural levels (*colonies and lamellae*) allows a consideration of them as hierarchically structured materials (HSM). Furthermore, an analogy is established in the paper between the microstructural arrangement in cold drawn pearlitic steels and the multi-level structure of Johann Sebastian Bach's music.

1. Prologue: *In the framework of "Don Quijote de la Mancha" by Miguel de Cervantes*

The high importance of wind energy in the field of the so-called *renewables* (or *clean*) energies is nowadays out of doubt. In this framework, wind energy turbines (Figure 1) are key structures to generate the energy. Although with different purpose, in certain sense they can be considered as engineering evolutions of the old-fashion windmills (Figure 2) appearing in the master work by Miguel de Cervantes Saavedra: "Don Quijote de la Mancha", whose cover page is reproduced in Figure 3 ("Don Quixote de la Mancha") in old Spanish, later evolved to modern Spanish, a developed and important language spoken nowadays by almost 550 million of people in the world. Figure 4 shows the Monument devoted to Miguel de Cervantes in the *Plaza de España* of Madrid, including not only the statue of the writer, but also of their creatures *Don Quijote de la Mancha* and *Sancho Panza*, whose profile is shown in Figure 5.

The main subject of this paper is related to fracture behaviour and structural integrity of cold drawn pearlitic steel wires for wind turbine structures and foundations. Therefore, such a structural steel will be considered as an energy material, since it is of key importance for the foundation (or even for the structure itself when it is made of prestressed concrete) of modern wind turbines. Special attention will be paid to the material-science type relationship between microstructure (including its geometrical arrangement) and macro- and microscopic fracture behavior.

IOP Conf. Series: Materials Science and Engineering **446** (2018) 012006 doi:10.1088/1757-899X/446/1/012006

Figure 1. Modern wind energy turbines.

Figure 2. Ancient windmills in Campo de Criptana (Ciudad Real, La Mancha, Spain).

EL INGENIOSO
HIDALGO DON QVI-
XOTE DE LA MANCHA,
Compueſto por Miguel de Ceruantes Saauedra.
DIRIGIDO AL DVQVE DE BEIAR,
Marques de Gibraleon, Conde de Benalcaçar, y Baña-
res, Vizconde de la Puebla de Alcozer, Señor de
las villas de Capilla, Curiel, y
Burguillos.

Año, 1605.

CON PRIVILEGIO,
EN MADRID Por Iuan de la Cueſta.
Vendeſe en caſa de Franciſco de Robles, librero del Rey nro ſeñor.

Figure 3. Cover page of "Don Quijote de la Mancha" by Miguel de Cervantes Saavedra.

Figure 4. Cervantes Monument in the *Plaza de España* (Madrid, Spain), including the statue of Miguel de Cervantes Saavedra and those of *Don Quijote de la Mancha* and *Sancho Panza*

Figure 5. Profile view of *Don Quijote* and *Sancho* (Cervantes Monument in Madrid).

2. Structural integrity of wind energy turbines

Structural health monitoring of wind turbine system is a critical issue in the development of wind energy [1,2]. In this framework, the two most important technical concerns are related to (i) the correct maintenance of the wind turbine gearbox [3-6], (ii) the structural integrity of the wind tower structure and foundation [7,8].

The author of the present paper has developed investigations in the specific field of damage tolerance of the wind turbine gearbox bearings [5,6], analyzing the phenomenon of hydrogen-assisted rolling-contact fatigue (HA-RCF) in the materials. In addition, his research is mainly concerned with fracture and structural integrity of eutectoid cold drawn prestressing steel wires to be used, e.g., in the main structure [7] and in the foundation or supporting system [8] of wind energy turbines in the case that prestressed concrete is the main structural material.

High-strength cold drawn pearlitic steel wires have extraordinary mechanical properties [9-12], so that they can be considered as *high-performance materials* (HPM) [9,10], their excellent behaviour being clearly influenced by the microstructural arrangement generated during the manufacture process by progressive (multi-step) cold drawing [10].

This paper offers a materials-science (micro- and macro-) approach to the fracture and structural integrity (in air and aggressive environments) of cold drawn pearlitic steel wires, by establishing a relationship between microstructure and fracture performance and introducing the innovative concept of microstructural integrity.

3. Hierarchical structure of cold drawn pearlitic steel wires

Cold drawn pearlitic steel wires are manufactured progressively in several steps (Figures 6 and 7) to increase the strength by a strain hardening mechanism, at the same time producing a microstructural evolution [13-15], as shown in Figure 8 in the form of slenderizing [16] and orientation [17] of the *first microstructural level* (parlitic colonies), together with densification [18] and orientation [19] of the *second microstructural level* (ferrite/cementite lamellae).

Figure 6. Manufacture of prestressing steel wires by progressive multi-step cold drawing.

Fracture of cementite lamellae or special interlamellar spacing appears in the research by Toribio *et al.* [20] who identified, denoted and described a new (*non-conventional*) microscrostructural unit: the *pearlitic pseudocolony* (Figure 9), a special pearlitic colony in which the lamellae are not oriented along the wire axis or cold drawing direction, thereby producing an anomalous (extremely high) pearlitic interlamellar spacing, sometimes associated with breaking (local fracture) and curling of cementite lamellae. These characteristics make them weakest areas or potential fracture initiation units, i.e., places able to produce fracture path deflection.

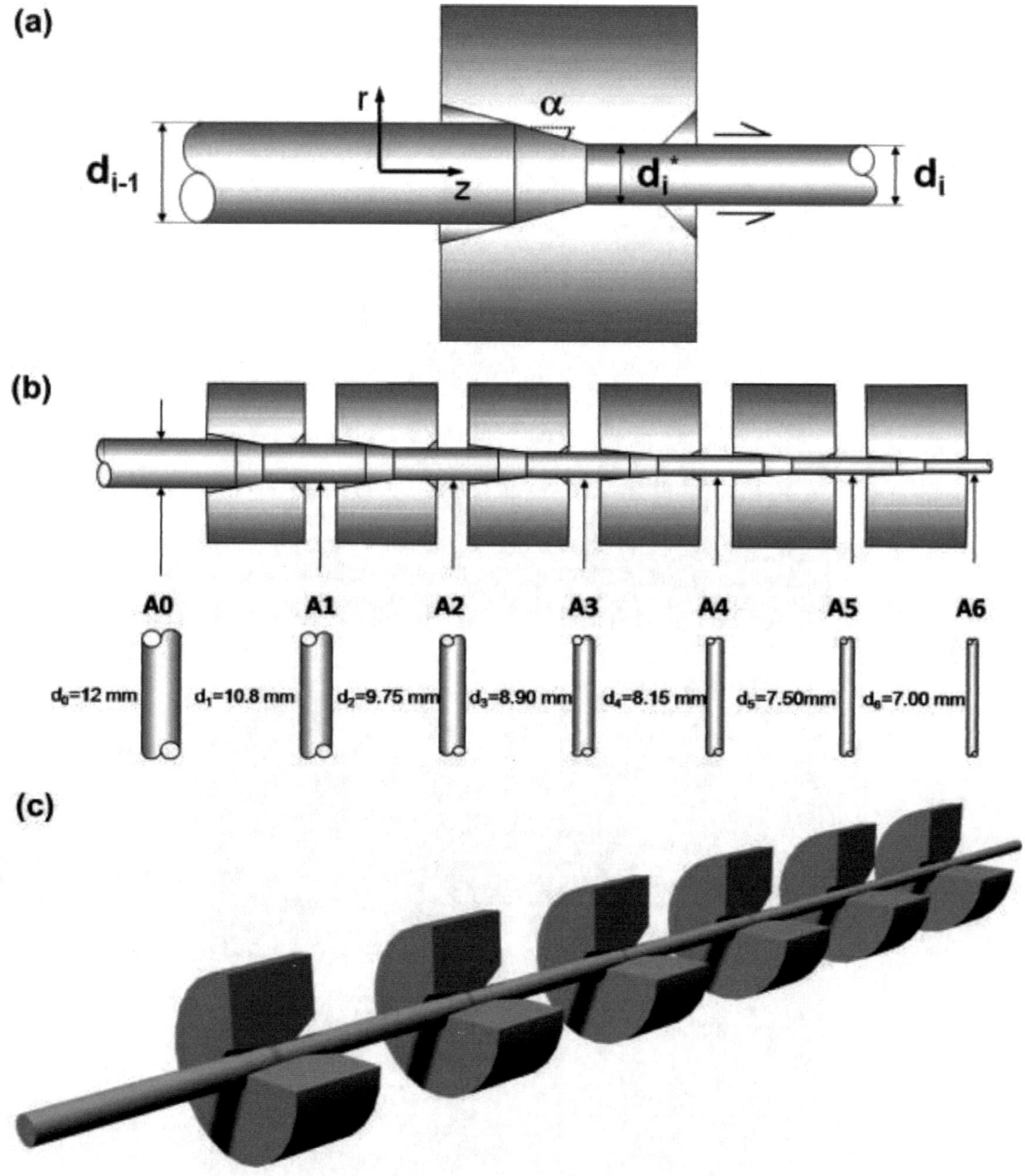

Figure 7. Scheme of the cold drawing process: (a) pass of the steel through the die; (b) progressive (repetitive) cold drawing in six passes; (c) 3D scheme of the whole manufacturing chain.

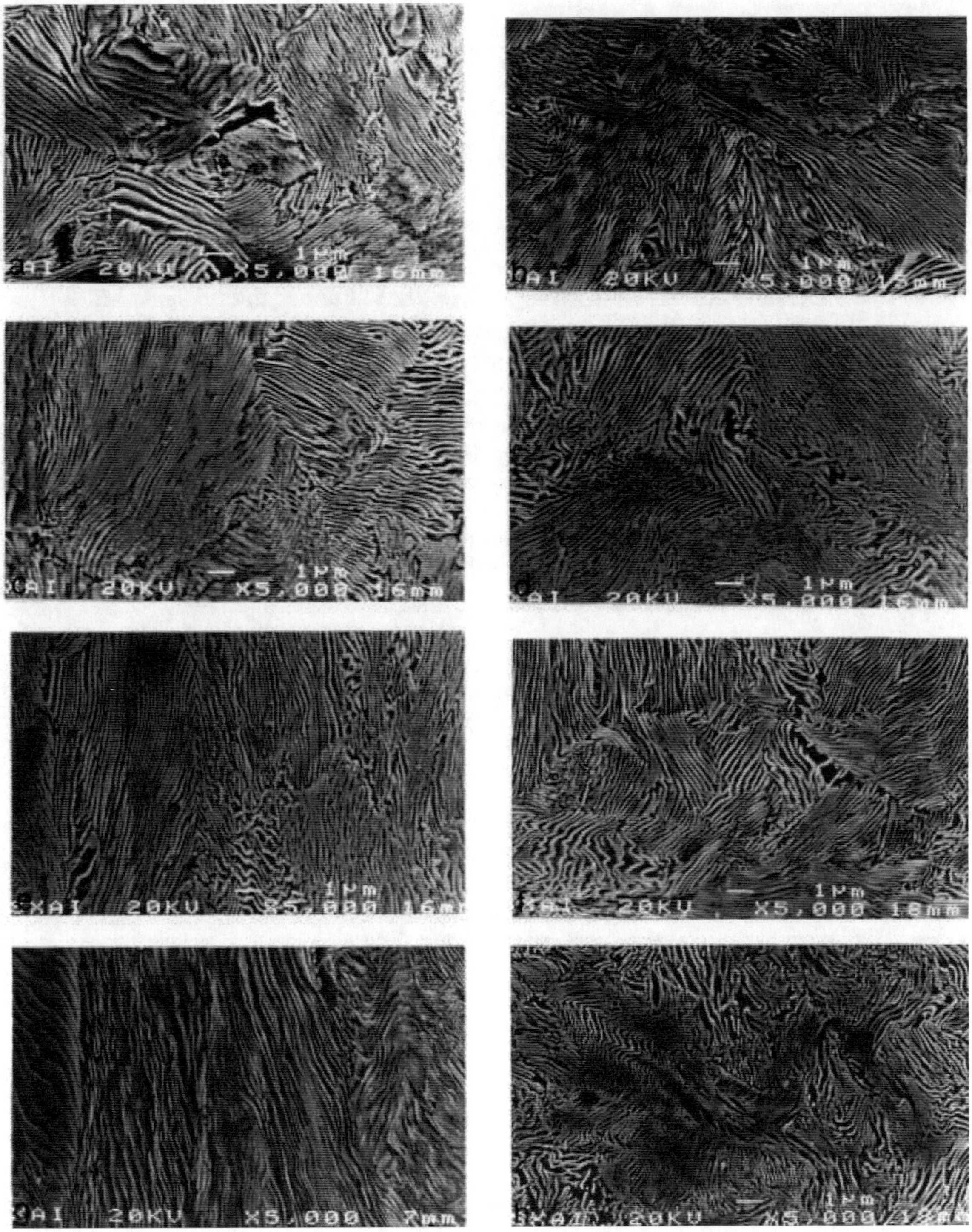

Figure 8. Microstructures of cold drawn pearlitic steels in longitudinal (left) and transverse (right) sections for steels that have undergone 0, 2, 4 and 6 steps of cold drawing (from top to bottom).

IOP Conf. Series: Materials Science and Engineering **446** (2018) 012006 doi:10.1088/1757-899X/446/1/012006

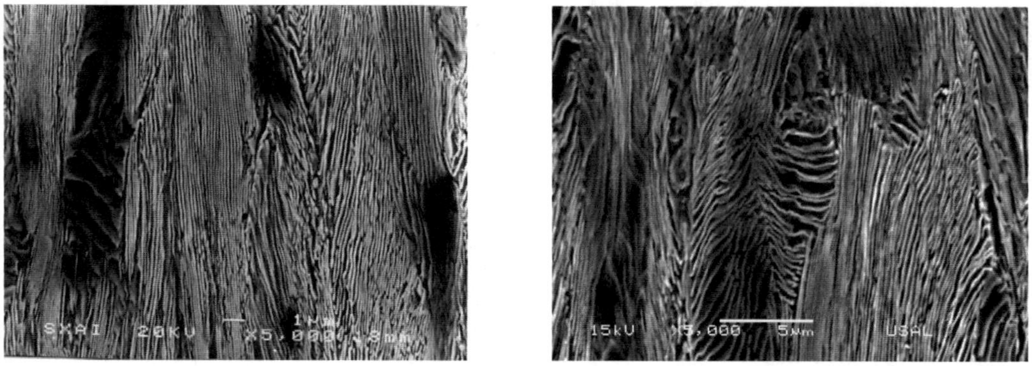

Figure 9. *Pearlitic pseudocolonies* in heavily cold drawn pearlitic steels after six drawing steps (longitudinal section).

The afore-said microstructural levels produce a hierarchical microstructure in the cold drawn pearlitic steels [16-21]. Therefore they can be regarded as *hierarchically-structured materials* (HSM) in which the different scales of analysis provide scientific information about the fracture performance in air and aggressive environments.

4. Fracture performance of cold drawn pearlitic steel wires

4.1. Fracture in air

Pearlitic steels with small (or null) degree of cold drawing fail in air (and at room temperature) in a predominantly ductile manner, so that the Miller & Smith micromechanical model of fracture in air by shear cracking in pearlite [22] has been proposed to explain the fracture process in slightly cold drawn pearlitic steel [23], as depicted in Figure 10. The plastic yielding of the tougher (an more ductile) ferrite leads to micro-fracture and yielding of the more brittle cementite lamellae.

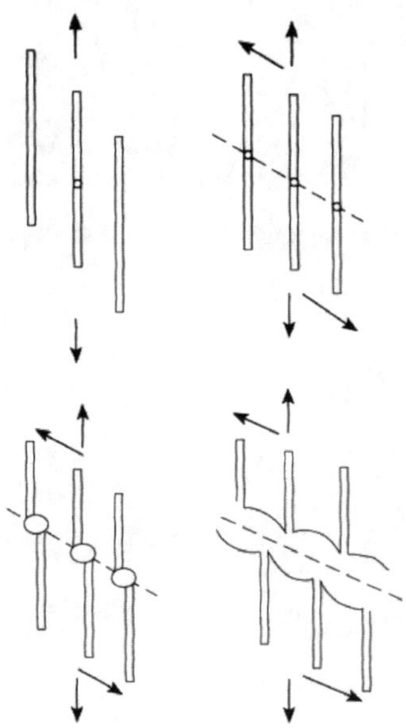

Figure 10. Microstructure-based model of fracture for slightly cold drawn pearlitic steel (on the basis of the Miller-Smith model of shear cracking of pearlite with fracture of cementite).

Figure 11 sketches the micromechanics of fracture in the case of heavily cold drawn pearlitic steel; as explained in previous paragraphs, the pseudocolonies represent weakest units promoting fracture and crack deflection appears when the macroscopic fracture path reaches a pearlitic pseudocolony, that itself fails by shear cracking of pearlite if not previously pre-cracked or pre-damaged as a consequence of the manufacture process by cold drawing [23], thus producing curling and micro-fractures.

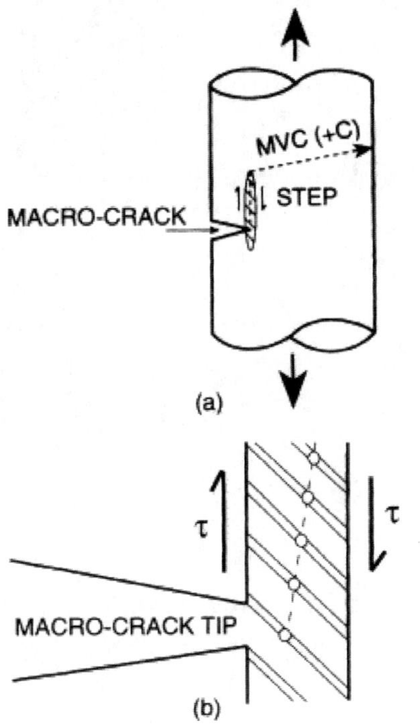

Figure 11. Microstructure-based model of fracture in the case of heavily cold drawn pearlitic steel (on the basis of the pearlitic pseudocolony as the weakest fracture unit promoting crack deflection).

4.2. Fracture in aggressive environments

The analysis of fracture of progressively cold drawn pearlitic steel was performed in previous research [24-26]. In the following sections the main results are described, with distinction between the two main regimes of environmental cracking.

4.2.1. Hydrogen assisted cracking (HAC)

Figure 12 shows microstructure-based models of fracture in progressively cold drawn pearlitic steels in the case of hydrogen assisted cracking (HAC). Depending on the degree of cold drawing, any of the two mechanisms of hydrogen embrittlement, *hydrogen enhanced localized plasticity* (HELP) or *hydrogen enhanced decohesion* (HEDE) could be operative, as described elsewhere [24].

For small or null degree of cold drawing, HELP is predominant, Figure 12(a), linked with shear cracking of pearlite and a micromechanical mode of fracture in the form of tearing topography surface or TTS, i.e., a mechanism with a non negligible degree of plasticity in the pearlitic microstructure, similar to that appearing in the case of fatigue crack propagation with transfer of the material from the crack tip to the flanks of the growing crack. This experimental fact could indicate the important role of plasticity in both fatigue crack propagation and HAC.

For heavily cold drawn pearlitic steels fractured in a hydrogen environment, HEDE is predominant, and in this case the concept of splitting or delamination fracture [26] is applicable. It could be decohesion, delamination or debonding between similar microstructural units, either the pearlitic colonies (the fracture path advancing just along the colony boundaries) or the ferrite/cementite lamellae, the delamination being through the interface, with a fracture mechanism similar to that of composite materials (as a matter of fact, the cold drawn pearlitic steel is a micro-composite material).

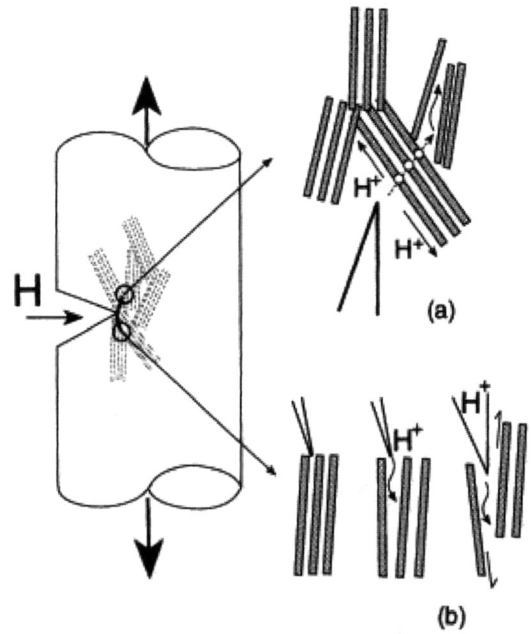

Figure 12. Microstructure-based model of hydrogen assisted cracking (HAC) in progressively cold drawn pearlitic steel.

4.2.2. Localized anodic dissolution (LAD)
Figure 13 shows a microstructure-based model of localized anodic dissolution (LAD) or pure stress corrosion cracking (SCC), cf. [25]. It could be valid for any degree of cold drawing.

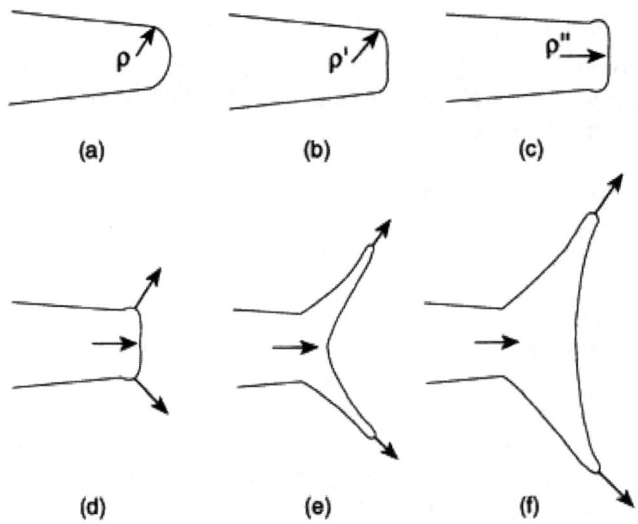

Figure 13. Microstructure-based model of localized anodic dissolution (LAD) in progressively cold drawn pearlitic steel

For pearlitic steels with low (or null degree of cold drawing), metal dissolution takes place and produces crack tip blunting (with round tip), cf. Figure 13(a). When the drawing level increases, the stronger anisotropy favors cornered blunting, cf. Figures 13(b) and 13(c) over round blunting. In the case of heavily cold drawn pearlitic steels, Figures 13(d), 13(e) and 13(f) show the sequence of enhanced dissolution just at the corners, finally producing crack deviation. In the most heavily drawn pearlitic steels, again the concept of splitting or delamination fracture [26] is applicable.

Apart from the described pioneering papers [23-26], dated more one decade ago, studying the relationship between hierarchical microstructure of cold drawn pearlitic steels and their fracture behavior, more recent papers present alternative analyses on the same topic [27-30]. In addition, the author of the present paper has published three very recent reviews [31-33] dealing with structural integrity [31], environmentally assisted cracking [32] and hydrogen effects [33] in progressively cold drawn pearlitic steel wires.

5. Closing remark about the relationship between microstructure and fracture behaviour

Cold drawing in pearlitic steel produces a *closely packed and oriented hierarchical* microstructure, thereby influencing the fracture behavior in air and aggressive environment which becomes very *anisotropic* (splitting or delamination) in the case of heavily cold drawn pearlitic steels.

6. Epilogue: *An approach to Johann Sebastian Bach*

On the basis of the hierarchical microstructure of cold drawn pearlitic steels (two levels of colonies and lamellae), a link can be established with the multi-scale (*multi-level or multi-layer*) structure of Johann Sebastian Bach's music. Figure 14 shows the Bach's image by Elias Gottlob Haussmann, the only authentic portrait of Johann Sebastian Bach (that in which the composer has the canon triplex).

Figure 14. Johann Sebastian Bach with the canon triplex (portrait by Elias Gottlob Haussmann).

In this framework, one really remarkable example is the Christmas Oratorio composed by six Cantatas (macro-structure), each of them with its own micro-structure (consisting of contrasting sections of recitatives, arias and choruses). Another beautiful evidence of multi-level structure of Bach's music is given in Figure 15 showing some early bars of the Fugue No 1 in C Major BWV 846 of The Well-Tempered Clavier (Book I). The (macro-) structure of the fugue itself contains (micro-) structure of thematic sections (*subject and counter-subject*) and connecting passages (*episodes or divertimenti*), i.e., it is a hierarchical structure like that of cold drawn pearlitic steel wires.

Figure 15. Fugue No 1 in C Major BWV 846 of The Well-Tempered Clavier (Book I).

References

[1] Ciang CC, Lee J-R and Bang H-J, Structural health monitoring for a wind turbine system: a review of damage detection methods, *Measurement Science and Technology* **19** (2008) 122001.

[2] Chen X, Li C and Tang J, Structural integrity of wind turbines impacted by tropical cyclones: A case study from China, *Journal of Physics: Conference Series* **753** (2016) 042003.

[3] Vegter RH and Slycke JT, The role of hydrogen on rolling contact fatigue response of rolling element bearings, *Journal of ASTM International* **7** (2010) 1-12.

[4] Evans M-H, Richardson AD, Wang L and Wood RJK, Serial sectioning investigation of butterfly and white etching crack (WEC) formation in wind turbine gearbox bearings, *Wear* **302** (2013) 1573-1582.

[5] Toribio J, Lorenzo M and Vergara D, Numerical analysis of hydrogen-assisted rolling-contact fatigue of wind turbine bearings, *Fracture and Structural Integrity* **30** (2014) 40-47.

[6] Toribio J, Lorenzo M and Vergara D, Role of multiaxial stress state in the hydrogen-assisted rolling-contact fatigue in bearings for wind turbines, *Fracture and Structural Integrity* **33** (2015) 434-443.

[7] Cortina-Cordero A, *Pre-stressed concrete tower for wind power generators*, US Patent Number US7739843B2 (2010).

[8] Lotfy I, *Prestressed Concrete Wind Turbine Supporting System*, MSc Thesis, University of Nebraska (2012)

[9] Toribio J, On the intrinsic character of the stress-strain curve of a prestressing steel, *Journal of Testing and Evaluation* **20** (1992) 357-362.

[10] Toribio J, Cold drawn eutectoid pearlitic steel wires as high performance materials in structural engineering, *Structural Durability and Health Monitoring* **2** (2006) 239-247.

[11] Borchers Ch and Kirchheim R, Cold-drawn pearlitic steel wires, *Progress in Material Science* **82** (2016) 405-444.

[12] Gil-Sevillano J, Cleavage-limited maximum strength of work-hardened B.C.C. polycrystals, *Acta Metallurgica* **34** (1986) 1473-1485.

[13] Embury JD and Fisher RM, The structure and properties of drawn pearlite, *Acta Metallurgica* **14** (1966) 147-159.

[14] Langford G, Deformation of pearlite, *Metallurgical Transactions* **8A** (1977) 861-875.

[15] Ridley N, A review of the data on the interlamellar spacing of pearlite. *Metallurgical Transactions* **15A** (1984) 1019-1036.

[16] Toribio J and Ovejero E, Microstructure evolution in a pearlitic steel subjected to progressive plastic deformation, *Materials Science and Engineering* **A234-236** (1997) 579-582.

[17] Toribio J and Ovejero E, Microstructure orientation in a pearlitic steel subjected to progressive plastic deformation, *Journal of Material Science Letters* **17** (1998) 1037-1040.

[18] Toribio J and Ovejero E, Effect of cumulative cold drawing on the pearlite interlamellar spacing in eutectoid steel, *Scripta Materialis* **39** (1998) 323-328.

[19] Toribio J and Ovejero E, Effect of cold drawing on microstructure and corrosion performance of high-strength steel, *Mechanics of Time-Dependent Materials* **1** (1998) 307-319.

[20] Toribio J Ovejero E and Toledano M, Microstructural bases of anisotropic fracture behaviour of heavily drawn steel, *International Journal of Fracture* **87** (1997) L83-L88.

[21] Zhang X, Godfrey A, Hansen N and Huang X, Hierarchical structures in cold-drawn pearlitic steel wire, *Acta Materialia* **61** (2013) 4898-4909.

[22] Miller LE and Smith GC, Tensile fractures in carbon steels. *Journal of the Iron & Steel Institute* **208** (1970) 998-1005.

[23] Toribio J, Microstructure-based modelling of fracture in progressively drawn pearlitic steels, *Engineering Fracture Mechanics* **71** (2004) 769-777.

[24] Toribio J and Ovejero E, Microstructure-based modelling of hydrogen assisted cracking in pearlitic steels, *Materials Science and Engineering* **A319-321** (2001) 540-543.

[25] Toribio J and Ovejero E, Microstructure-based modelling of localized anodic dissolution in pearlitic steels, *Materials Science and Engineering* **A319-321** (2001) 308-311.

[26] Toribio J, Delamination fracture of prestressing steel: an engineering approach, *Engineering Fracture Mechanics* **75** (2008) 2683-2694.

[27] Guo N, Luan B and Liu Q, Influence of pre-torsion deformation on microstructures and properties of cold drawing pearlitic steel wires, *Materials & Design* **50** (2013) 285-292.

[28] Zhao T, Zhang G, Zhang S and Zhang L, Influence of lamellar direction in pearlitic steel wire on mechanical properties and microstructure evolution, *Journal of Iron and Steel Research International* **23** (2016) 1290-1296.

[29] Zhao T, Zhang G, Zhang S and Zhang L, Fracture characteristics of fully pearlitic steel wire in tension and torsion, *Journal of Iron and Steel Research International* **23** (2016) 1206-1212.

[30] He Y, Xiang S, Shi W, Liu J, Ji X and Yu W, Effect of microstructure evolution on anisotropic fracture behaviors of cold drawing pearlitic steels *Materials Science and Engineering* **A683** (2017) 153-163.

[31] Toribio J, Structural integrity of progressively cold-drawn pearlitic steels: *From Rafaello Sanzio to Vincent van Gogh*, *Procedia Structural Integrity* **3** (2017) 3-10.

[32] Toribio J, Stress corrosion cracking of progressively cold-drawn pearlitic steels: *From Tintoretto to Picasso*, *Procedia Structural Integrity* **5** (2017) 1439-1445.

[33] Toribio J, Hydrogen effects on progressively cold-drawn pearlitic steels: *Between Donatello and Michelangelo*, *Procedia Structural Integrity* **5** (2017) 1446-1453.

Acknowledgements
This paper corresponds to the Keynote Plenary Lecture delivered in the Conference ICEMA 2018 (3rd *International Conference on Energy Materials and Applications*) held in the Campus Viriato of the University of Salamanca, Zamora, Spain, the year 2018 during which the University of Salamanca celebrates its 800th anniversary, it being the oldest University in Spain and one of the five oldest in the

world (logos of the University of Salamanca are shown below these lines). The author wants to thank Emma Wang and Lily Chen, Chengdu Young Education & Consultancy Co., Ltd, for their management of the conference and for their kindness.

In memoriam

The paper is dedicated to the memory of Professor Dr. Robert Goldstein, master and friend, who passed away in 2017.

IOP Conf. Series: Materials Science and Engineering **446** (2018) 012007 doi:10.1088/1757-899X/446/1/012007

An extended hot plate method for measurement of thermal conductivity variation with temperature of building materials

Saad Raefat[1,b], Mohammed Garoum[1,a,*], Najma Laaroussi[1], El Mehdi El Khattabi[2], Mohammed Rhachi[1]

[1] Materials, Energy and Acoustics TEAM, Higher School of Technology Salé, Mohammed V University in Rabat, 11060, 227 Avenue Prince Héritier, Morocco.
[2]Laboratory for Integration of Advanced Systems and Technologies-University Sidi Mohamed ben Abdellah-Faculty of Sciences Dhar El Mehraz, Fez, Morocco.

agaroum1@yahoo.fr,bsaad.raefat@gmail.com

Abstract: A new exploitation of the Hot Plate in Steady state HPS on the determination of thermal conductivity variation with temperature is presented. The mean temperature changement with temperature of the tested sample is ensured with varying the value of the voltage applied to the hot plate resistance, which enables to measure the thermal conductivity in different temperatures. Calculated thermal conductivity values obtained using a gypsum plaster sample, which specific heat capacity variation with temperature is accurately measured, validate the technic. Results show that the mean calculated thermal conductivity of the studied gypsum plaster is 0,143 W/m.k while the measured one is 0,148 W/m.k with a mean deviation of 3,81%. Moreover, the two trend lines of measured values and calculated ones have the same slope with a small intercept difference of 0,004.

1. Introduction

At lower temperatures, two classifications of experimental methods of the thermal conductivity determination exist: steady state and transient state methods. In the first category the sample thermal conductivity is directly calculated measuring the flow density and the temperature gradient once the system falls into equilibrium [1, 2]. In the second category the thermal diffusivity is measured instead, which takes into account the temperature fluctuations inside the sample, the thermal conductivity is then estimated from the thermal diffusivity knowing the specific heat and the density of the sample [3]. The hot plate method [4] is considered to be a first category technic for thermal conductivity measurements of small samples (0,01m^2). In steady state, its straightforward elements incorporation made its use simple and easy, while keeping an acceptable accuracy of measurements. In the present study this technic use is extended to the determination of the thermal conductivity variation with temperature of construction materials. A gypsum plaster sample was used to present and explain the experimental and numerical procedures. The specific heat variation with temperature of gypsum plaster was measured using an accurate differential scanning calorimeter in order to compare calculated and measured values of the thermal conductivity assuming a constant thermal diffusivity and density.

2. Description of used material

IOP Conf. Series: Materials Science and Engineering **446** (2018) 012007 doi:10.1088/1757-899X/446/1/012007

The sample was made in a circular mold of 10cm in diameter and 2cm in thickness mixing 190g of gypsum plaster with 118,75g of water respecting a water ratio of 0,625 as indicated in NF P 15-201-1, it was left on an ambient temperature of 25 ° C and a RH of 80% for a sufficient time to consistency, subsequently introduced into an oven and dried at a temperature of 70 ° C + -3 ° C according to NF P 75-101. Once the sample mass becomes constant it is immediately coated on cellophane in order to keep its dry mass constant until measures.

3. Experimental approach:

3.1.Description of the stationary hot plate method

Figure 1 describes the experiment of the Hot Plate in Steady state HPS, the heating element is composed of a plane heating resistance of dimensions given in Table 1 inserted between the sample and the polyethylene having thermal conductivity of λ_p = 0,043 $W.m^{-1}K^{-1}$. Most of the heat dissipated by the heating element of a resistance R_h= 101Ω passes through its upper part, the different temperatures are measured by a thermocouple type K with a resolution of 0.001K. The set is then placed between two aluminium blocks in order to ensure a constant temperature on the unheated faces of the sample and the polyethylene foam in one hand, in the other hand to underwrite a faster establishment of the steady state, the three presented thermocouples measure the temperatures T_0, T_1, T_2 which correspond to temperatures measured; in the centre of the lower face of the heating element, on the unheated face of the sample and on the unheated side of the polyethylene, respectively.
Admitting the explained configuration one can write in steady state and supposing that the transfer by conduction is on one-dimensional:

$$\phi_0 = \phi_1 + \phi_2 \tag{1}$$

ϕ_0, ϕ_1 and ϕ_2 are expressed as:

$$\phi_0 = \frac{U.I}{S} \tag{2}$$

$$\phi_1 = \frac{\lambda_{plaster}}{e_{plaster}} \times (T_0 - T_1) \tag{3}$$

$$\phi_2 = \frac{\lambda_p}{e_p} \times (T_0 - T_2) \tag{4}$$

The voltage U applied through the heating element is estimated at an accuracy of 0.01 V.

By combining the previous equations we can write that:

$$\lambda_{plaster} = \frac{e_{plaster}}{T_0 - T_1} \times \left[\frac{U.I}{S} - \frac{\lambda_p}{e_p} \times (T_0 - T_2) \right] \tag{5}$$

Table 1. Dimensions of the elements of the circular HPS

Element	Heating element	polyethylene	Aluminium block
Dimensions (Diameter, thickness)	100x002 mm	100x10 mm	100x55 mm

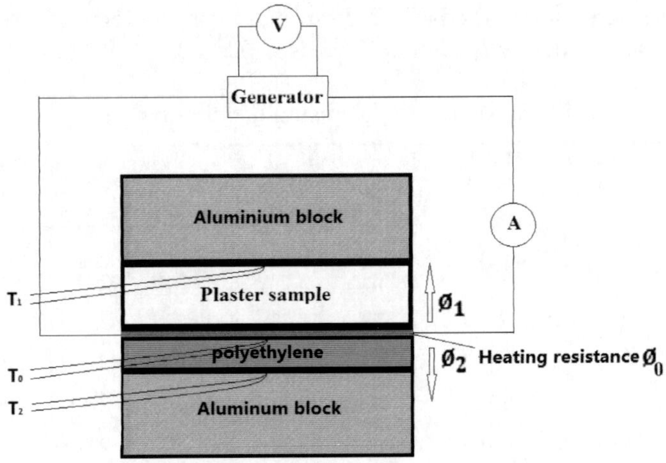

Figure 1: Schematic diagram of the circular HPS.

3.2. Using the HPS to determine the thermal conductivity variation

In this study the HPS elements are having a circular form; it is inserted in an insulated box in order to ensure a temperature stabilisation during the experimentation. In order to guarantee an acceptable accuracy the circular HPS (CHPS) is relied to a voltmeter and an ammeter to measure respectively the tension and the current as indicated in Figure 1. From equation (5) we noticed that the thermal conductivity, which corresponds to a mean temperature $T_m = (T_0 + T_1)/2$ of the sample, is proportional to the electric tension applied to the heating resistance. In order to evaluate the evolution of the thermal conductivity versus temperature the tension U is changed in the generator respecting a voltage step equal to 1V. For high mean temperatures T_m and for each value of U the thermal conductivity is measured once the steady state is established. However for low T_m four ice accumulators are disposed around the circular HPS as illustrated in Figure 2, which enables to measure thermal conductivity in low temperatures.

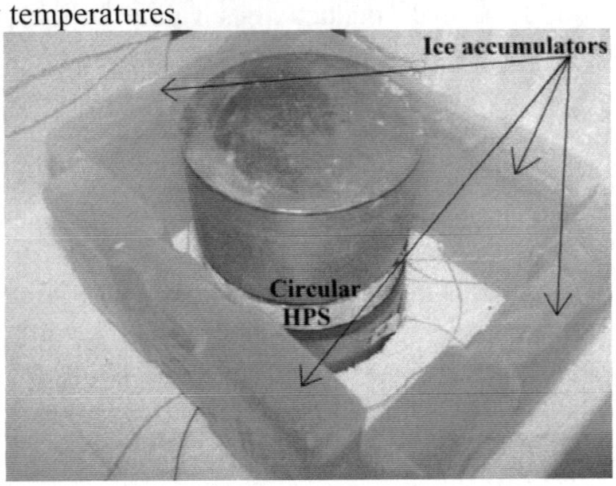

Figure 2: The circular HPS disposition in low temperatures measurements.

4. Results and discussion

4.1. Determination of thermal conductivity variation with temperature

Table 2 presents the measured thermal conductivity of plaster in the temperature range [19°C; 48°C], it is clear that the thermal conductivity slightly increases from 0,146 W/m.K for 19,1 °C to 0,154°C for 47,7 °C.

Table 2. Measured data using the circular HPS.

Circular HPS Disposition	U (V)	I (A)	T_0 (°C)	T_1 (°C)	T_2 (°C)	T_m (°C)	$\lambda_{plaster}$ (W/m.K)
With ice accumlators	12,08	0,119	26,148	11,584	12,058	19,103	0,146
	13,08	0,129	29,225	12,049	12,569	20,897	0,145
	14,09	0,139	32,678	12,780	13,224	22,951	0,144
	15,08	0,149	36,292	13,513	13,853	25,072	0,144
	16,08	0,159	39,749	14,111	14,267	27,008	0,146
	17,08	0,169	44,004	15,242	15,174	29,589	0,147
Without ice accumlators	10,08	0,100	36,316	26,213	26,082	31,199	0,145
	11,08	0,109	38,847	26,782	26,616	32,732	0,145
	12,08	0,119	41,895	27,628	27,462	34,679	0,147
	13,08	0,129	44,687	28,185	27,978	36,333	0,150
	14,09	0,139	47,986	28,907	28,692	38,339	0,151
	15,08	0,149	51,700	29,732	29,642	40,671	0,151
	16,08	0,159	55,324	30,366	30,332	42,828	0,152
	17,08	0,169	59,63	31,578	31,062	45,346	0,151
	19,08	0,189	65,133	30,364	30,205	47,669	0,154

4.2. Validation of the thermal conductivity variation

In order to validate the measured thermal conductivities values in the previous sub-section, other values were calculated knowing:

- The density: Calculated knowing the sample mass and apparent dimensions at dry state. The density is equal to 1002,909 kg/m^3.
- The thermal diffusivity: measured using the flash technic [5] in a mean temperature of the sample 28°C and calculated adopting the global minimization procedures detailed in [6]. The calculated thermal diffusivity is 1,767 10^{-7} m^2/s.
- The specific thermal capacity: measured using the differential scanning calorimeter µDSC evo7 designed for the study of samples in isothermal modes over a wide temperature range [-45°C; 120°C], in this study the specific heat is measured in the temperature range [5°C; 50°C], the values of C_p corresponding to the studied temperatures in CHPS were extracted from the measures in this temperature range and collected in Table 3.

Table 3. Measured specific thermal capacity of gypsum plaster using the µDSC evo7

Tm (°C)	Cp (J.kg-1.K-1)	Tm (°C)	Cp (J.kg-1.K-1)	Tm (°C)	Cp (J.kg-1.K-1)
19,103	782,287	29,589	798,178	38,339	812,942
20,897	785,679	31,199	801,329	40,671	816,865
22,951	788,561	32,7315	803,463	42,828	820,828
25,0725	791,691	34,6785	806,943	45,346	824,533
27,008	794,014	36,3325	809,904	47,669	827,57

The thermal conductivity variation is then calculated using equation (6) and assuming that the thermal diffusivity and density remain constant versus temperature.

$$\lambda(T) = \rho \times a \times c(T) \tag{6}$$

Figure 2 presumes the thermal conductivity distribution of both measured and calculated values of thermal conductivity and the deviation percentage between them. It is shown that both values are having a positive trend with the same line slope 3.10^{-4} and a small difference of the intercept of 0,004. The deviation, measured using equation (7), goes from 1,83% to 5,29% with a mean value of 3,81%. These results validate the current exploitation of the HPS.

$$\text{Deviation}(\%) = \frac{|\lambda_{\text{calculated}} - \lambda_{\text{measured}}|}{\lambda_{\text{calculated}}} \times 100 \tag{7}$$

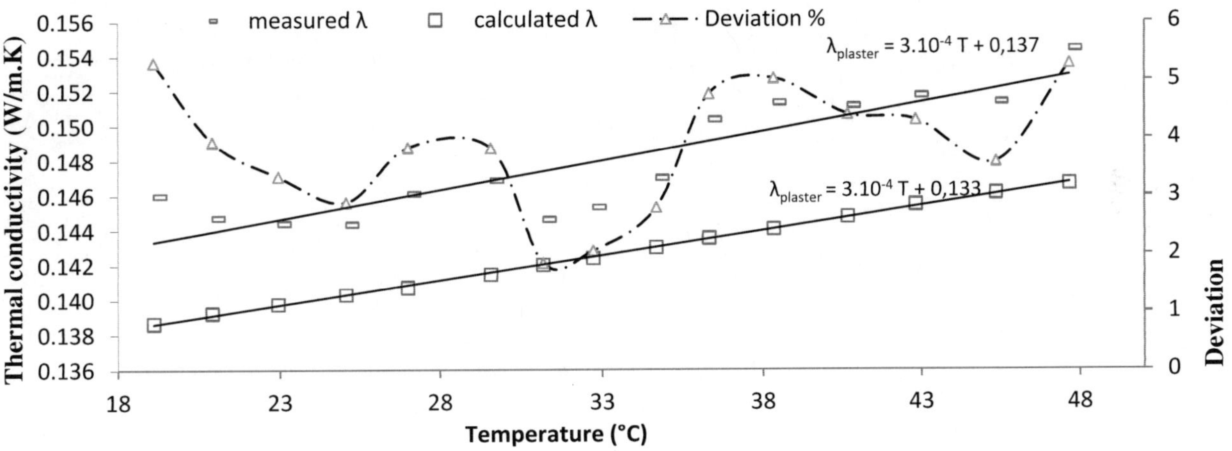

Figure 3: calculated and measured values of Gypsum plaster thermal conductivity versus temperature.

5. Conclusion

A straightforward exploitation of the hot plate method allowing the measurement of thermal conductivity variation with temperature of construction materials has been presented. The main interest of this study stays in its capability to perform a series of measurements in steady state ensuring a fast establishment of the thermal conductivity distribution versus temperature while conserving the same state of the sample. An experiment has been conducted on a sample of gypsum plaster enabling to compare its values with calculated ones; adopting experimental values of the specific heat variation with temperature and assuming a constant thermal diffusivity and density. We have given in this study the resulting deviations between calculated and measured values of thermal conductivity variation in the temperature range of [19°C; 48°C], this range is mostly located in interior temperatures of Moroccan buildings. A larger range could be studied using a heat source inside the box which contains the CHPS.

References:

[1] White, G. K., & Woods, S. B. Electrical and thermal resistivity of the transition elements at low temperatures. Philosophical Transactions of the Royal Society of London A 1959 *Mathematical, Physical and Engineering Sciences* **251(995)** 273-302.

[2] White, G. K. Measurement of Solid Conductors at Low Temperatures in Thermal Conductivity 1969 **vol. 1** chap. 2.

[3] Biot, J. B. Traité de physique expérimentale et mathématique 1816 **Vol. 1**.

IOP Conf. Series: Materials Science and Engineering **446** (2018) 012007 doi:10.1088/1757-899X/446/1/012007

[4] Jannot, Y., Remy, B., & Degiovanni, A. Measurement of thermal conductivity and thermal resistance with a tiny hot plate2009 *High Temperatures-High Pressures* **39** 11-31.
[5] Parker, W. J., Jenkins, R. J., Butler, C. P., & Abbott, G. L. Flash method of determining thermal diffusivity, heat capacity, and thermal 1961 conductivity *Journal of applied physics* ***32*(9)** 1679-1684.
[6] Raefat, S., Garoum, M., Laaroussi, N., Thiam, M., & Amarray, K. Thermal diffusivity and adiabatic limit temperature characterization of consolidate granular expanded perlite using the flash method 2017 *IOP Conference Series: Materials Science and Engineering* **Vol. 222**, No. 1, 012004.

Numerical inverse estimation of the thermal diffusivity and the adiabatic limit temperature of three types of unfired clay bricks using flash method and global minimization algorithm

Yassine Chihab[a1], Saad Raefat[a], Mohammed Garoum[a2], Najma Laaroussi[a], Rachid Bouferra[b]

aMohammed V University in Rabat, EST de Salé, Materials, Energy and Acoustics TEAM, Morocco.
[b]Cadi Ayyad University, Marrakech, Morocco.

[1]yassinechihab@research.emi.ac.ma ; [2]garoum1@yahoo.fr

Abstract. Clay is an ecological building material used widely in constructions especially in rural areas. In this work an experimental study has been carried out to estimate the apparent thermal diffusivity and adiabatic limit temperature of three types of unfired clay brick using flash method. The three samples were prepared with the same water proportion (picked between the plasticity and the liquidity limits) and exposed to the same drying conditions. The thermal diffusivity and adiabatic limit temperature were estimated using numerical global minimization procedures of the quadratic distance between the theoretical model and the experimental thermogram. The theoretical model is derived from a numerical inversion of Laplace transform using Gaver-Stehfest method. An approach based on the independence between the sensitivity coefficients and the magnitude of these coefficients(Beck criterion)is presented in order to examine the simultaneous estimation of the different parameters related to the theoretical model from the experimental data. For all samples tested, the results show that the estimated parameters give a good agreement between the experimental data and the theoretical model.

1. Introduction

The building sector accounts for 28% of energy consumption and contributes about a third of the CO_2 emissions. In Morocco, the building sector consumes more than 25% of final energy [1]. So, the reducing of the energy cost of building has become a necessity. Unfired Clay is an abundant raw natural resource and it is one of the oldest construction materials needing a very low amount of energy during the manufacturing process.

Many studies have been published concerning the thermal behavior of material based on clay. El Azhary et al. [2] studied the impact of using unfired clay on the energy efficiency of building in the south of Morocco. The thermal properties of fired clay were measured by Laaroussi et al.[3]. Another work of El Azhary et al.[4] studied the effect of mass fraction variation of straw on the thermal properties of composite material clay-straw.

This work is focused on the numerical estimation of the thermal diffusivity and the adiabatic limit temperature for three different unfired clay bricks using the flash method.

2. Experimental approach

2.1 Samples preparation

Three different raw clays extracted from three Moroccan sites (clay of Marrakech (CM), clay of Tamansourt (CT) and clay of Essaouira (CE)) were studied. Figure 1 shows the grain size distribution while the plasticity (PL), the liquidity (LL)limits and the plasticity index (PI) are reported in the Table 1.
Three samples were then prepared using the same water proportion. The water to clay ratio was kept equal to 28%.The samples were then dried in an oven at 70°C during three days and packed in plastic films to avoid any moisture contamination.

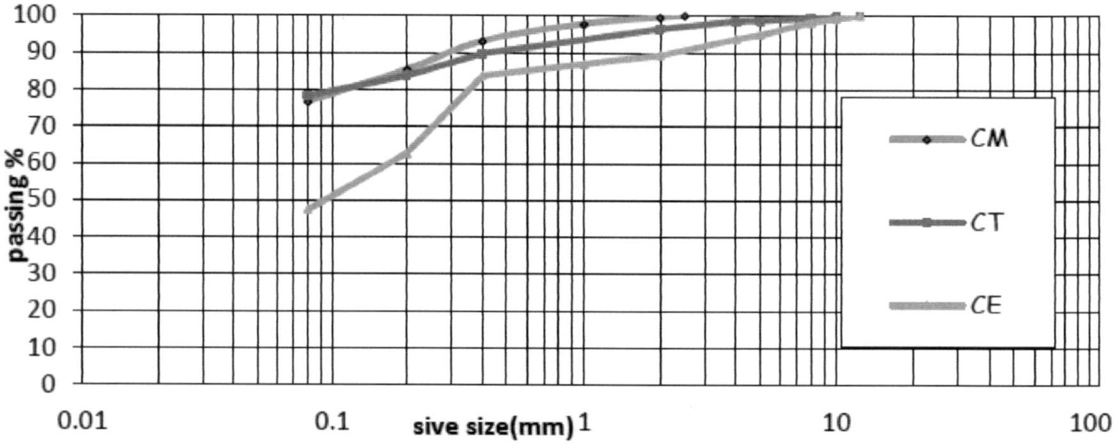

Figure 1.Grain size curves of the three clays

Table 1.Atterberg limits of the three raw clays used

sample	Atterberg Limits		
	Liquidity Limit(%)	Plasticity Limit(%)	Plasticity Index(%)
Clay of Marrakech (CM)	37	23	14
Clay of Tamansourt (CT)	65	25	40
Clay of Essaouira (CE)	46	26	20

2.2 The flash method
This method is mainly used to estimate the thermal diffusivity of materials. Its basic principle is described in he Figure 3. The front face of the sample is exposed to a strong heat pulse and the variation of the rise emperature with time is measured at the rear face using a thermocouple type K.

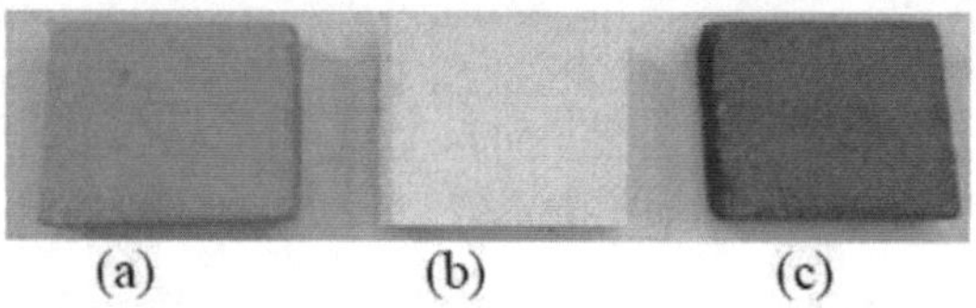

Figure 2. Studied samples (a) CT, (b) CE, (c) CM

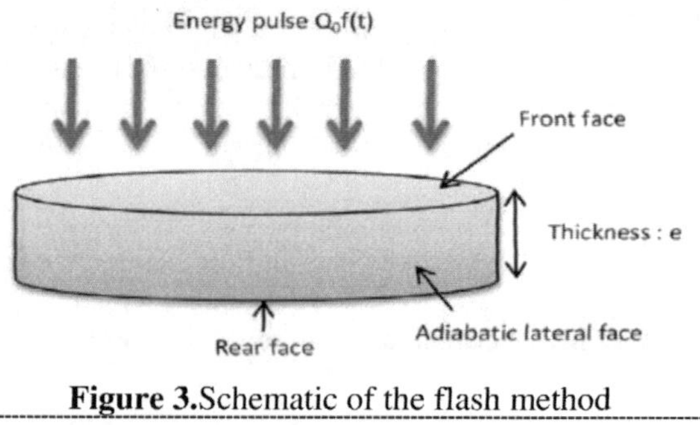

Figure 3.Schematic of the flash method

3. Mathematical model

The front face of samples was covered by a thin black painting to havecomplete absorption of the radiative flux.We assume that the absorbed energy is uniform and theinitial temperature$T_0$of the sample isequal to the ambient one. In order to allow a one dimension heat transfer across the sample, itslateral faceswere insulated.The rise temperature T isthengoverned by the following system(1).

$$\begin{cases} \dfrac{\partial^2 T(x,t)}{\partial x^2} = \dfrac{1}{a}\dfrac{\partial T}{\partial t}. \\ -\lambda.\dfrac{\partial T(x,t)}{\partial x}\Big|_{x=0} = -h_1.T(0,t) + Q_0.f(t). \\ -\lambda.\dfrac{\partial T(x,t)}{\partial x}\Big|_{x=e} = h_2.T(e,t). \\ f(t) = \begin{cases} \dfrac{1}{t_d} & , \quad 0 \le t \le t_d. \\ 0 & , \quad t_d \le t. \end{cases} \\ T(x,0) = 0. \end{cases} \tag{1}$$

Where a is the thermal diffusivity, h_1 and h_2 are the global heat exchange coefficients on both sides of the sample, $[Q_0.\times f(t)]$ is the finite pulse with duration t_d, $Q_0(W/m^2)$is the finite amount of heat absorbed at the front boundary (x=0) and f(t) is the time dependence of the heat generation assimilated as a square function. For all measurements, the duration of flash t_d is kept equal to 4s.

The Laplace transform of Eq.1allows expressing the Laplace transform θ of the rise temperature as:

$$\theta(e,T_{ma},a,b_{i1},b_{i2},p) = T_{ma}.F(p)\frac{e^2}{a}.\frac{1}{\left[\dfrac{(b_{i1}.b_{i2})A(p)}{\sqrt{\dfrac{e^2}{a}p}} + (b_{i1}+b_{i2})B(p) + A(p)\sqrt{\dfrac{e^2}{a}p}\right]}. \tag{2}$$

With: $A(p) = \mathrm{Sinh}\left(\sqrt{\frac{e^2}{a}p}\right)$ and $B(p) = \mathrm{Cosh}\left(\sqrt{\frac{e^2}{a}p}\right)$.

e is the thickness of the sample, b_{i1} and b_{i2} are the Biot numbers respectively, $b_{i1} = \frac{h_1 e}{\lambda}$ and $b_{i2} = \frac{h_2 e}{\lambda}$

$F(p) = \frac{1-e^{-p.t_d}}{p.t_d}$. is the Laplace transform of .f (t)

$T_{ma} = \frac{Q_0}{\rho c e}$ is the adiabatic limit temperature. Where c is the specific heat and ρ is the density.

The theoretical thermogram of the rise temperature on the rear face $T_{th}(e,T_{ma},a,b_{i1},b_{i2},t)$is then obtained by inversing numerically the solution (2) using the Gaver-Stehfest method [5].

$$T_{th}(e,T_{ma},a,b_{i1},b_{i2},t) = \frac{\ln(2)}{t}\sum_{j=1}^{n} V_j.\theta\left(e,T_{ma},a,b_{i1},b_{i2},\frac{\ln(2)}{t}.j\right) \tag{3}$$

The coefficients $V_j$are given by the relation:

$$V_j = (-1)^{(\frac{n}{2}+j)} \sum_{k=\left[\frac{j+1}{2}\right]}^{Min(j,\frac{n}{2})} \frac{(k)^{(\frac{n}{2}+1)}(2k)!}{(\frac{n}{2}-k)!(k)!(i-k)!(2k-1)!} \tag{4}$$

The even number n of terms used in the summation must be optimized. Increasing n increases the precision of the result to a certain limit, and then the precision decreases due to the increase in rounding errors. The Gaver-Stehfest method was chosen for its swiftness and its easier numerical implementation. A special script was written in Mathematica language.[6].

4. Local sensitivity analysis and global minimization

The local sensitivityanalysis permits to know how a small disturbance of the different parametersinput ofthe theoretical model T_{th}influencesthe variabilityof its output. The relative coefficient of sensitivity$\chi_k$is defined as:

$$\chi_k = k\frac{\partial F}{\partial k}$$

For this model $F=\frac{\ln(2)}{t}\sum_{j=1}^{n} V_j.\theta\left(e, T_{ma}, a, b_{i1}, b_{i2}, \frac{\ln(2)}{t}.j\right)$ and $k=\eta_k$

$$\chi_k = \eta_k \frac{\ln(2)}{t}.\sum_{j=1}^{n} V_j.\frac{\partial\theta\left(e, T_{ma}, a, b_{i1}, b_{i2}, \frac{\ln(2)}{t}.j\right)}{\partial\eta_k}, 1 \le k \le 4 \tag{5}$$

Where η_k is unknown parameters which be estimated with (η_1=a, η_2=T_{ma} ,η_3=b_{i1},η_4=b_{i2}). According to Beck criterion [7], the sensitivity coefficients must be linearly independent. In addition χ_k must be different fromzero. The analysis of Sensitivity coefficient permits to know if all parameters can be simultaneously estimated from the experimental data with a good precision.

Figure 4 shows an example of relative coefficients of sensitivity curves for the theoretical model T_{th}.

It is noticed that all sensitivity coefficients are different from zero, The analysis of the Figure 4 shows that for low time range (t<300s) this model is very sensitive to the change in *a* and T_{ma} on the other hand it is very weakly sensitive to the change in b_{i1} and b_{i2}.For time range (t≥300s), this model becomes sensitive to the parameters b_{i1} and b_{i2} but the sensitivity coefficients related to these parameters are linearly dependent. We can explain this Interrelationship by the symmetry of existing model for b_{i1} and b_{i2}. In this time range, all parameters are correlated and the theoretical model doesn't respect the predictability Beck's criterion.

This model is very weakly sensitive to the change in b_{i1} and b_{i2}, consequently the adopted estimation approach can conduct to several values of these parameters. Therefore, they can't be determined numerically with a good precision from the experimental data. In addition, the sensitivity coefficients related to *a* and T_{ma} meets the beck criterion for low time range (t<300s), so these tow parameters can be estimated numerically with a good precision in this time range.

However, the four parameters T_{ma}, *a* b_{i1} and b_{i2}were simultaneously estimated by minimizing the quadratic distance M between T_{th} and the experimental thermograms.

$$M(e, T_{ma}, a, b_{i1}, b_{i2}) = \sum_{(j=1)}^{N}\left[T_{exp}(t_j) - T_{th}(e, T_{ma}, a, b_{i1}, b_{i2}, t_j)\right]^2. \tag{6}$$

$t_j$is the time points and N is the length of the experimental data vector taken into account.

The minimization of the Eq. 6 of the multimodal function M is solved using Nelder Mead algorithm [8].

5. Results and discussion

5.1 Density:

From the knowledge of the dimensions and masses of the dry samples, we can easily deduce their apparent densities which are presented in Table 2.It is clear that the heaviest types of unfired clay is (CM) with density equal to 1890.256 kg/m^3. The clay (CT) and (CE) have almost the same density.

Table 2. Predicted thermal diffusivity, adiabatic limit temperature and samples density

Sample	Test	Estimated diffusivity $a\times10^{-7}[m^2.s^{-1}]$	Measurement deviation (%)	Estimated T_{ma} [°C]	Measurement deviation (%)	ρ [kg/m^3]
CM	1	3.660	1.723	0.964	1.903	1890.256
	2	3.625	0.750	0.909	3.911	
	3	3.509	2.474	0.965	2.008	
	Mean value	**3.598**		**0.946**		
CT	1	3.285	1.139	1.034	2.783	1616.83
	2	3.264	0.492	1.000	0.596	
	3	3.197	1.570	0.985	2.087	
	Mean value	**3.248**		**1.006**		
CE	1	3.004	0.033	1.047	0.852	1616.16
	2	3.007	0.066	1.041	1.420	
	3	3.006	0.033	1.081	2.367	
	Mean value	**3.005**		**1.056**		

5.2 Thermal diffusivity and adiabatic limit temperature:

From the Table 2, it is clear that the thermal diffusivity decreases with the decrease of density, the (CM) has the greatest diffusivity with values equal to $3.598\times10^{-7}m^2/s$on other hand the (CE)has the lowest ones. According to the results of Table 2, it is noticed that the adiabatic limit temperature increases with the decrease of density. In addition, the(CE) has the highest value of the adiabatic limit temperature which means that it has the lowest value of volumetric capacity (ρ.c).

Figures (5,6,7) show that the estimated parameters for all samples tested give a good agreement between the experimental data and the theoretical model.

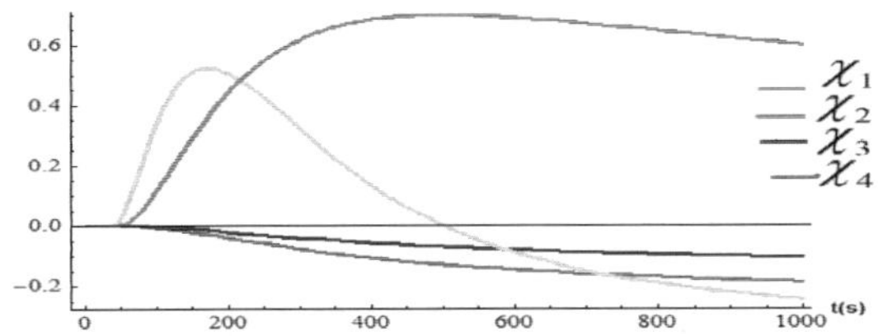

Figure 4. Relative coefficients sensitivity curves(b_{i1}=0.2 b_{i2}=0.1 e=0.02m a=$3\times10^{-7}m^2/s$ T_{ma}=1°C)

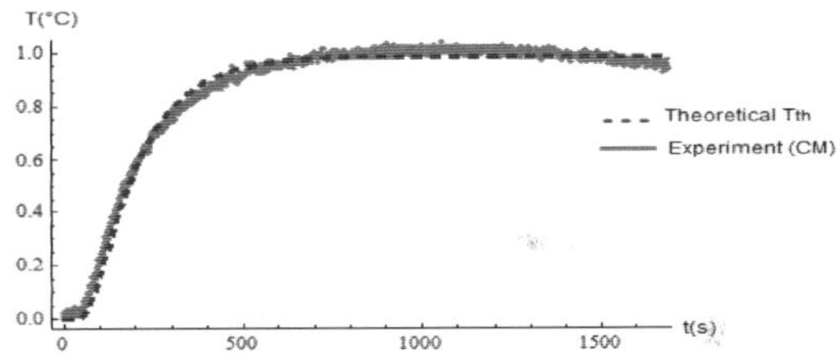

Figure 5. Comparison between the theoretical and experimental data(CM)

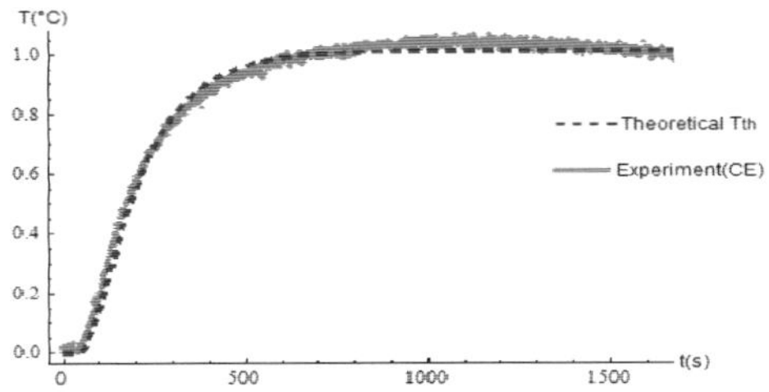

Figure 6. Comparison between the theoretical and experimental data(CE)

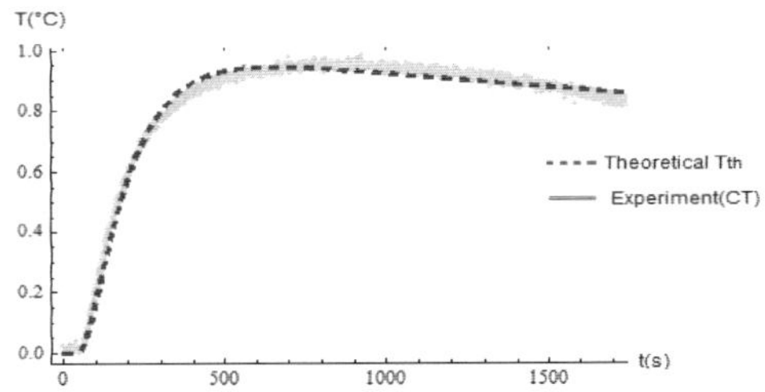

Figure 7. Comparison between the theoretical and experimental data(CT)

6. Conclusion

This work has presented an estimation of the diffusivity and the adiabatic limit temperature of three types of unfired clay samples using the flash method combined with a global minimization procedure of the quadratic distance between the theoretical model and the experimental thermograms. From the results, it can be concluded that the thermal diffusivity increases with increase of density while the adiabatic limit temperature decreases with increase of density. It is showed that the estimated parameters for all tested samples conduct to a good agreement between the experimental data and the theoretical model.

References

[1] Moroccan Agency for the Energy Efficiency (AMEE).Règlement Thermique de Construction au Maroc. 2014. www.amee.ma.
[2] El Azhary. K.. Lamrani. A.. Raefat. S.. Laaroussi. N.. Garoum. M.. Mansour. M..& Khalfaoui. M. (2017). The improving energy efficiency using unfired clay envelope of housing construction in the south Morocco. Journal of Materials and Environmental Sciences. 8. 3771-3776.
[3] Laaroussi. N.. Lauriat.G.. Garoum.M..Cherki. A..& Jannot.Y. (2014). Measurement of thermal properties of brick materials based on clay mixtures. Construction and Building Materials. 70. 351-361.
[4] El Azhary. K.. Chihab.Y..Mansour. M.. Laaroussi.N..& Garoum.M. (2017). Energy Efficiency and Thermal Properties of the Composite Material Clay-straw. Energy Procedia. 141. 160-164.
[5] Stehfest. Numerical inversion of Laplace transforms.Communications of the Association for Computing Machinery. 1970.
[6] Raefat, S., Garoum, M., Laaroussi, N., Thiam, M., & Amarray, K. (2017, July). Thermal diffusivity and adiabatic limit temperature characterization of consolidate granular expanded perlite using the flash method. In IOP Conference Series: Materials Science and Engineering (Vol. 222, No. 1, p. 012004). IOP Publishing.
[7] Beck.J.V. Parameters estimation in engineering and science. Wiley& sons. USA. 1989.
[8] J. A. Nelder and R. Mead, A simplex method for function minimization, Computer Journal 7 (1965), 308–31.

IOP Conf. Series: Materials Science and Engineering **446** (2018) 012009 doi:10.1088/1757-899X/446/1/012009

Research on energy storage system participating in frequency regulation

Huating Jiang[1,a],Lijun Qin[1]
[1] Electric Engineering, North China Electric Power University, 102206, Beijing,China
E-mail:ajhuating@126.com

Abstract. With increasing penetration of renewable source in power system, higher requirements for power quality are put forward. Energy storage system represented by chemical battery and flywheel energy storage system is fast-ramping and responses quickly in frequency regulation market. It shows outstanding performance in frequency regulation comparing with the traditional frequency regulation resource. This paper reports a review of the energy storage system participating in frequency regulation, including frequency regulation market and energy storage technology. Also, it contrasts the frequency regulation characteristics and total costs between battery energy storage system (BESS) and flywheel energy storage system (FESS) both applied widely in the projects. The operation mode and Simulink modelling of energy storage system, along with the control strategy and capacity configuration, are also discussed through relative literature.

1. Introduction

Energy storage system has a wide range of applications in the smart grid, especially plays an important role in frequency regulation market. The energy storage system has variety kinds, including the traditional chemical battery, pumped storage, but also advanced flywheel energy storage, compressed air energy storage, supercapacitors energy storage and superconducting energy storage. How to achieve large-scale, secure and stable storage of electric energy is a critical issue. With the wind power, solar energy and other renewable energy connecting to the grid, it has more requirements for power quality.

At present, the main frequency regulation resources are thermal power units and pumped storage. While the thermal power units have to bear the task of adjusting the power peak. In the heating period, it also needs to bear the heating task. The location of pumped storage power station is limited by geographical location, water head, terrain and geology, etc. Compared with the conventional frequency regulation resources (FRR), energy storage resources such as chemical storage, flywheel energy storage, response and regulate quickly with a high accuracy. As the cost of batteries reduces and the frequency regulation compensation rises, the economic benefits of large-scale energy storage are even more substantial participating in the frequency regulation market.

Recently, many countries have attached great importance to energy storage technologies and applications. Xtreme Power, A123 and other companies have invested demonstration projects involving a variety of energy storage types, including lithium-ion batteries, ranging from 1MW/0.5MWh to 20MW/5MWh [1] [2]. Beacon Power participates in 20MW flywheel storage station for frequency regulation in Hazle, Unite States. From multiple sources [3] [4], the energy storage plants have been built mainly involved in frequency regulation in recent years. These projects shows it is potential that energy storage participating the frequency regulation.

2. Research Status Quo

2.1. Frequency regulation market

The United States has a relatively complete frequency regulation market. From 2007 to 2013, FERC issued Order 890, 755, 784 and 792 to guarantee the legitimacy of energy storage system participating in frequency regulation market and obtain reasonable compensation [5]. In particular, Order 755(2011) presents a new compensation method [6]. It compensates not only for the frequency regulation capacity, but also the frequency mileage and performance. By this way, the energy storage system can acquire more reasonable compensation because of their fast-ramping characteristic. Order 792 defines energy storage as small power equipment and has developed a rapid interconnection test. Besides, standard IEEE 1547 is applied for physical and electrical interconnections between utility and distributed generation.

Chinese frequency regulation market is undergoing reform. At present, there are no explicit standards about the energy storage system participating in the frequency regulation market, usually referring to the standards about the generator or generation management methods in the power market. The energy storage system mostly installed in the power plant is also part of generators' AGC assessment. AGC assessment of generators can be based on "the two rules". On June 7, 2016, the National Energy Administration issued the notice on "Pilot Project of Electric Power Storage Ancillary Service Compensation Mechanism (Market) to promote the Participation of Electric Storage Energy", and started pilot projects about the energy storage system used for frequency regulation [7]. In November 2017, the National Energy Administration issued the "Work Plan for Perfecting Compensation Mechanism (Market) for Ancillary Services"[8]. These measures are conducive to energy storage system participating in the frequency regulation and energy storage system will receive more reasonable compensation.

2.2. Energy storage technologies

At present, the main frequency regulation resources are pumped storage and thermal power; Fast-ramping resources include chemical batteries, flywheel energy storage, superconducting energy storage, supercapacitors and compressed air storage. Among them, the superconducting energy storage, supercapacitors and compressed air technology are not mature enough to apply in projects. The comparison between FESS and BESS are as follows.

2.2.1. Characteristics comparison.

From the aspects below, we can learn the differentials between FESS and BESS.

- Depth of discharge (DOD): When the DOD is deep, such as 80%~100%, FESS can maintain its charge and discharge characteristics, while the lithium ion cannot. And the cycle number of lithium ion will decrease to a large extent.
- Temperature: FESS and lithium batteries have different operating temperatures. Lithium iron phosphate battery, for example, the charging temperature is 0 ~ 45 ℃, discharge temperature is -20 ~ 75 ℃ [9]. In the operating temperature, temperature changes will not affect the performance of flywheel energy storage, but will affect the performance of lithium batteries. Lithium battery charge and discharge characteristics, self-discharge rate, depth of discharge are all affected by temperature.
- Service life: FESS's design life is 20 years, while the lithium-ion battery 4 years. It is an advantage over the BESS.
- Discharge time: Flywheel energy storage battery is a high-power energy storage device, so that it cannot maintain long-term discharge. However, the lithium batteries are high-capacity device and can discharge for a relatively long time.
- Safety and environmental impact: FESS is safe, environment-friendly without any chemical substance. Lithium batteries have some security risks, which should be equipped with fire alarm and other related equipment.

IOP Conf. Series: Materials Science and Engineering **446** (2018) 012009 doi:10.1088/1757-899X/446/1/012009

The comparison between FESS and BESS can list in Table 1 below.

Table 1.FESS and BESS characteristic comparison

	FESS	BESS
Response time	Full cacacity within seconds	Fast response result in degradation
Accurate output	Fast	Fast
Charge or dischage	None	Avoid over discharging or over charging
Depth of discharge(DOD)	Permit deep discharge	Avoid deep discharge
degration	None	Degrate with cycles
cycles(100%DOD)	>100000	<2500
life	15-20year	About 4 years(it depends)
safety	Safe	Unsafe
Environmental inpact	Environmental friendly， no hazardous chemicals	chemicals

2.2.2. Cost comparison.

The cost of energy storage system mainly includes the initial investment cost, operation and maintenance costs, replacement costs, recycling costs, the flywheel energy storage battery and lithium battery cost are as follows[10].

- The initial investment cost: When the capacity is certain, the battery has a lower initial investment cost, flywheel energy storage battery is about more than 1.5-2.0 times the initial investment cost of lithium batteries.
- Operation and maintenance costs: the energy storage system has to pay a lot for the operation and maintenance for the better characteristics. In contrast, flywheel energy storage operation and maintenance costs lower.
- Replacement costs: flywheel energy storage life is about 15-20years, it need no replacement cost during its service life. While a lithium battery is about 4years (it depends on specific kind).It means that it replaces every 4years to maintain battery availability.
- The cost of recycling: lithium batteries need to be recycled due to the existence of chemical substances, and flywheel energy storage needn't. As the flywheel storage system is mechanical structure without chemical substance.

Based on the different performance of frequency regulation and costs of FESS and BESS, it is important to choose the kind and certain capacity of frequency regulation resource.

3. Energy storage operation mode and Simulink Modelling

Energy storage operation mode can be divided into two kinds: independent energy storage frequency regulation station and joint operation mode. The flywheel storage frequency regulation station in Stephen is an example of independent operation mode. Joint operation mode requires energy storage system assist the traditional frequency regulation resource, such as thermal power units, hydroelectric units, photovoltaic, and wind power. According to the number of energy storage type, it can be divided into single energy storage and mixed storage operation mode. Energy storage systems usually assist the traditional thermal power plant to improve the quality of frequency in China.

The energy storage system is composed of energy storage system and Power Conversion System (PCS). The energy storage device is connected to the grid via a DC / DC, DC / AC, filter, and transformer [11], as shown in Fig. 1.

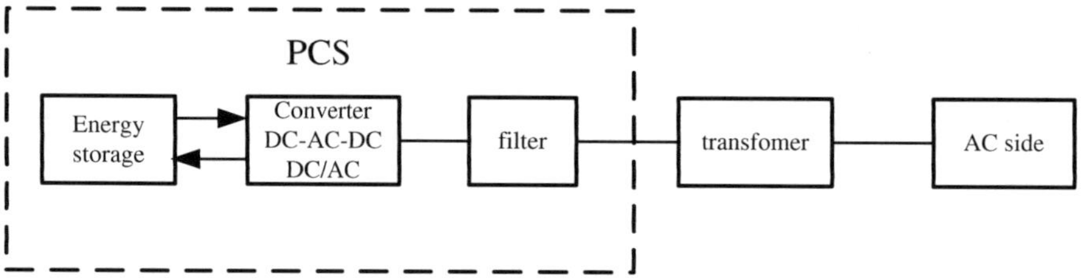

Figure1. The Energy Storage System Diagram

In terms of energy storage, taking BESS as an example, its model mainly includes a single element model and a PCS with a control method [12]. In the frequency regulation field, the product of the one-order inertial loop and the control gain is widely used to describe the behaviour of the BESS or FESS involved in the grid frequency regulation [13]. Besides, BESS can also described by external characteristic equivalent circuit [14].

4. Control strategy and capacity configuration

The control strategy of energy storage involved in the frequency regulation is to determine the control mode, timing and depth of action. Based on the regional grid, the ACE or ARR is calculated and then allocated. Literature [15] proposes that the discrete Fourier transform (DFT) for the change of frequency regulation demand is decomposed into low frequency and high frequency components. The energy storage system acts on high frequency components. Literature [16] compares the frequency regulation effect, energy storage state under the control of ARR and ACE. Energy storage system is priority or proportionate strategy. The literature [14] prioritizes energy storage systems (electric vehicles) participating in the AGC in Danish power system. The literature [17-18] distributes the ratio of the dynamic optimization between the flywheel and hydro-power unit based on AGC signal. Consider the thermal power units and energy storage device frequency regulation different internal characteristics to choose different distribution based on different scenarios. Literature [19] presents an optimal control strategy allocation considering suppression of reverse frequency regulation, dead-zone oscillation, over-discharge of energy storage device in the frequency regulation system composed of thermal power, VRB and VRLA.

Capacity configuration is to determine the degree of energy storage involved in the frequency regulation, that is, the capacity of energy storage and power. For independent or joint operation of energy storage frequency regulation station capacity configuration flow can describe as six steps:1) Get regional power grid frequency regulation historical data and control signals to determine the frequency regulation demand.2) Choose the energy storage kind based on the frequency regulation characteristics and conditions.3) Deploy the energy storage capacity and power.4) Find the constrains (technology: SOC, economy: costs and frequency regulation compensation).5) Analysis the results 6) Determine the optimal capacity. If the results in 5) cannot demand, it should return to the 2) or 3) until find the optimal capacity.

5. Conclusion

In order to promote the application of energy storage system in the frequency regulation market, we can promote it from the following aspects:

- A sound frequency regulation market, policies and compensation mechanisms are the preconditions for energy storage to participate in the frequency regulation market. The PJM market is a good example for energy storage system participating in frequency regulation market. The technical development and cost reduction can also promote.
- Discuss the modelling and operation mode of the energy storage system to determine the optimal objective function and configure the optimal capacity of the energy storage system.

References

[1] CHEN Dayu, ZHANG Lizi, WANG Shu, et al. Development of energy storage in frequency regulation market of United States and its enlightenment[J].Automation of Electric Power Systems, 2013, 37(1): 9-13.

[2] ZHANG Chuan, YANG Lei, NIU Tongyang, et al. Comparison and analysis of energy storage technology to balance fluctuation of wind power output [J]. Power System Protection and Control, 2015, 43(7): 149-154.

[3] LI Xinran, HUANG Jiyuan, CHEN Yuanyang, et al. Review on large-scale involvement of energy storage in power grid fast frequency regulation [J]. Power System Protection and Control, 2016, 44(7):145-153.

[4] LI Xiangjun, WANG Shanghang, HUI Dong. Summary and Prospect of Operation Control and Application Method for Battery Energy Storage Systems [J]. Power System Technology, 2017(10):3315-3325.

[5] FERC: Major Orders & Regulations [EB/OL]https://www.ferc.gov/legal/maj-ord-reg.asp

[6] FERC Order No.755: frequency regulation compensation in the organized wholesale power markets[EB/OL].http://www.ferc.gov/whats-new/comm-meet20111020ll/E-28.pdf

[7] CHEN Dapeng ,JING Zhaoxia. Analysis of Frequency Regulation Compensation Mechanism in Frequency Regulation Auxiliary Service Market [J]. Automation of Electric Power Systems, 2017(18):1-9.

[8] Government Information Disclosure Directory --- National Energy Administration --- Notice of National Energy Administration on Printing and Distributing the Work Plan of Perfecting Compensation Mechanism for Electric Power Auxiliary Services (Market) Mechanism[EB/OL] http://zfxxgk.nea.gov.cn/auto92/201711/t20171122_3058.htm

[9] Luo Haifeng. Study on Intelligent Management System of LiFePO4 Battery[D].Wuhan University of Technology, 2013.

[10] Energy storage benefits and market analysis handbook [R].America: Sandia National Laboratories, 2009.

[11] HUANG Jiyuan , LI Xinran, HUANG Jijun, et al. Comparison of Different Types of Energy Storage Power Supplies in Power System Frequency Modulation[J] Advanced Technology of Electrical Engineering and Energy, 2015(3):49-53.

[12] CAO Shengyun, SONG Chunning, LIN Xiaofeng, et al. Study of PCS's control strategy for battery energy storage grid-connected system[J]. Power System Protection and Control, 2014, 42(24): 93-98.

[13] MU Yunfei, WU Jianzhong, EKANAYAKE J, et al. Primary frequency response from electric vehicles in the great Britain power system[J]. IEEE Transactions on Smart Grid, 2013, 4(2): 1142-1150.

[14] J R Pillai , B Bak-Jensen . Integration of vehicle-to-grid in the western Danish power system ［J］. IEEE Transactions on Sustainable Energy, 2011, 2 (1) : 12-19.

[15] HU Zechun, XIE Xu, ZHANG Fang, et al. Research on Automatic Generation Control Strategy Incorporating Energy Storage Resources [J]. Proceedings of the CSEE, 2014, 34(29):5080-5087.

[16] HUANG Yawei, LI Xinran, HUANG Jiyuan, et al. Analysis of Control Methods for AGC with Battery Energy Storage System [J]. Proceedings of the CSU-EPSA, 2017, 29(3):83-89.

[17] Nyeng P, Yang B, Ma J, et al. Coordinated multi-objective control of regulating resources in multi-area power systems with large penetration of wind power generation[C]//7th International Workshop on Large-Scale Integration of Wind Power into Power Systems. Madrid, Spain, 2008.

[18] JIN Chunlian, LU Ning, LU Shuai, et al. A coordinating algorithm for dispatching regulation services between sloand fast power regulating resources [J]. IEEE Trans on Smart Grid, 2014, 5 (2) : 1043-1050.

[19] NIU Yang , ZHANG Feng ,ZHANG Hui, et al. Hybrid Energy Storage Optimal Control and Capacity Planning to Improve AGC Performance of Thermal Power Units [J]. Automation of Electric Power Systems, 2016, 40(10):38-45.

IOP Conf. Series: Materials Science and Engineering **446** (2018) 012010 doi:10.1088/1757-899X/446/1/012010

On the optimization of InGaP/GaAs/InGaAs triple-junction solar cell

Omar Saif[1,2], Mohamed Abouelatta[2], Ahmed Shaker[2], M K Elsaid[2]

[1] Al-Madina Higher Institute for Engineering and Technology, Giza, Egypt
[2] Faculty of Engineering- Ain Shams University, Cairo11517, Egypt.
Email: o.saif@hotmail.com

Abstract. This paper presents an optimization procedure for the design parameters of InGaP/GaAs/InGaAs Triple-Junction (TJ) solar cell by using SILVACO TCAD. The solar cell design parameters include layers' materials, thicknesses and doping concentrations. Firstly, the optimization technique is performed on an InGaP/GaAs/Ge cell. The Ge sub-cell is then replaced by an InGaAs sub-cell. A comparison between the performance parameters of InGaP/GaAs/InGaAs and InGaP/GaAs/Ge TJ solar cells is investigated. The compared parameters are the open circuit voltage (V_{OC}), short circuit current density (J_{SC}), Fill Factor (FF) and the conversion efficiency (η). Finally, a comparison between these optimized devices against some previously published work is presented. All simulations for triple-junction solar cells are accomplished under light intensity of 1-sun of standard AM1.5G solar spectrum at 300 K. The electrical characteristics for the proposed TJ solar cell are V_{OC} = 2.9 V and J_{SC} = 15.97 mA/cm^2 with conversion efficiency = 42.01%.

1. Introduction

Monolithic multi-junction (MJ) solar cells, fabricated from III-V compound semiconductors, have been developed in order to achieve high power conversion efficiencies for terrestrial and space applications [1–3]. An MJ solar cell is a combination of different types of photovoltaic junctions; each junction forms a sub-cell, stacked over one another via homo-junctions, intrinsic materials or tunnel junctions. Each sub-cell absorbs sunlight at a specific wavelength region of the solar spectrum and converts it into electricity. Moreover, matching the current generated by each sub-cell is crucial to obtain a maximum power from the solar cell. MJ solar cells are capable of generating a power approximately twice the power generated by the conventional solar cells [4].

An Experimental efficiency of around 41.6% was obtained by Boeing Spectrolab Inc. with a state-of-art TJ GaInP/GaInAs/Ge solar cell under 484 suns, AM1.5D and 25°C. Furthermore, high efficiency inverted metamorphic (IMM) InGaP/GaAs/Ge multi-junction solar cells are developed as the next space solar cell architecture. Spectrolab's achieved an efficiency of 33% at 1-sun and AM0 [5,6].

There were many publications that introduced studies of MJ solar cells and compared their simulation results with the experimental results. The earlier designs for TJ devices were based on GaAs and Ge substrates due to the well-known electronic and optical properties of the lattice-matched hetero-structures [7]. Remarkable progress in this field led to efficiencies higher than 34.5% under 1-sun global illumination and higher than 40.6% under concentrated direct sunlight [8].

The rest of the paper is organized as follows: In section 2, the design and simulation of the TJ solar cells are introduced. In section 3, the optimization for InGaP/GaAs/InGaAs TJ solar cell is performed.

IOP Conf. Series: Materials Science and Engineering **446** (2018) 012010 doi:10.1088/1757-899X/446/1/012010

Then, the simulation results are compared with some previously published results in section 4. Finally, section 5 is devoted to identifying paper's conclusions.

2. TJ solar cell design

2.1. Device structure

The InGaP/GaAs/InGaAs triple-junction solar cell structure shown in figure 1 consists of three p-n junctions and two n^+GaAs/p^+GaAs tunnel junctions. The top InGaP and middle GaAs cells, each consists of a window, p-n junction, Back Surface Field (BSF) layer and Buffer layer. The InGaAs material is added as epitaxial layers to form the bottom cell. Real-like doping concentration levels are used. InGaP and GaAs doping concentrations never exceed 5×10^{19} cm^{-3} [9] while InGaAs doping concentration of 3×10^{18} cm^{-3} is selected. To design an optimized solar cell, the window must be chosen carefully to be lattice-matched with that of the emitter layer, having a wide energy bandgap compared to the emitter and providing a potential barrier for minority carrier of the emitter. In this work, AlGaAs is used as a top window. The BSF layer is selected in order to passivate the surface recombination between the base of sub-cell and the emitter of the tunnel diode. So, the BSF must be lattice-matched to the GaAs and has a higher energy bandgap than the base of the bottom cell. The InGaP layer is selected to be the BSF. The tunnel junction is provided between two sub-cells in order to avoid the performance degradation of the MJ solar cells [10].

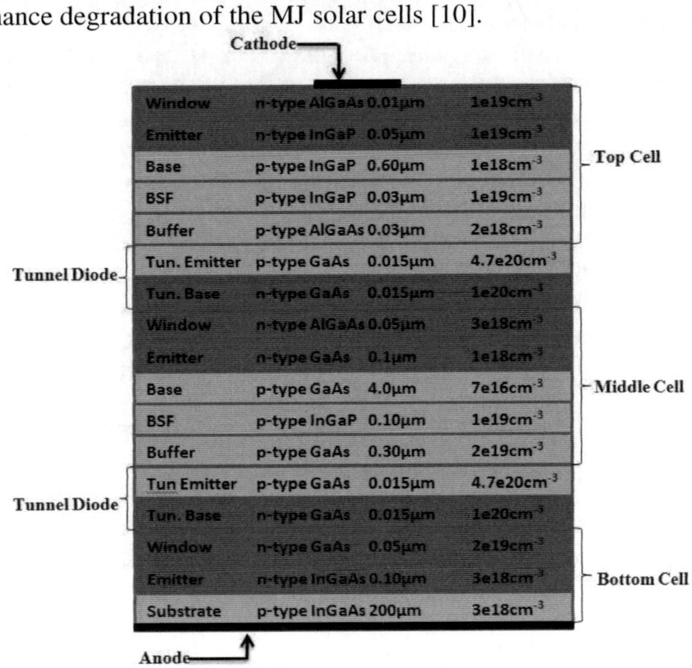

Figure 1. Optimized InGaP/GaAs/InGaAs TJ solar cell structure used in this design indicating the design parameters.

2.2. Simulation of InGaP/GaAs/Ge triple-junction solar cell

The TJ solar cell device is simulated using SILVACO ATLAS® tool and the performance of the cell is evaluated under 1-sun of AM1.5G standard spectrum. The physical models are reported as follows. To model the doping-dependent low-field mobilities of holes and electrons, the Concentration-Dependent Low Field Mobility (CONMOB) is used. Regarding recombination, Optical Recombination (OPTR) and Shockley-Read-Hall (SRH) recombination models are enabled. SRH recombination model considers the electrons being emitted or captured by donor and acceptor-like traps. The OPTR determines the possibility that a photon is generated when an electron and hole recombine. To model the non-local band-to-band tunneling, (BBT.NONLOCAL) model is selected [11]. As well, LUMINOUS is an ATLAS module which determines the photo-generation at each mesh point of the generated structure. Extinction coefficients as well as refractive index of each material are important parameters for the simulation of solar cells. Both extinction coefficients as well as refractive index are obtained from [12-14].

Figure 2 shows the simulated I-V characteristics for the InGaP top cell, the GaAs middle cell, the Ge bottom cell and the InGaP/GaAs/Ge TJ solar cell under 1-sun of AM1.5G illumination. The simulation results show that the open circuit voltages for InGaP, GaAs and Ge sub-cells are 1.5 V, 1.06 V and 0.35 V, respectively. Moreover, the InGaP cell produces the least current among the three individual sub-cells, which is 15.5 mA/cm^2, while GaAs and Ge sub-cells produce 30.78 mA/cm^2 and 59.74 mA/cm^2, respectively. As well, the InGaP/GaAs/Ge TJ solar cell has the following parameters, V_{OC} = 2.76 V and J_{SC} = 15.5 mA/cm^2. It is clear that the value for V_{OC} of TJ solar cell is approximately, the sum of V_{OC} of the three sub-cells, while the current of the TJ solar cell is determined by the sub-cell which has the lowest current density among the three sub-cells i.e. the top cell.

2.3. InGaP/GaAs/InGaAs triple-junction solar cell Design and Simulation

In this section, the Ge bottom cell of TJ solar cell is replaced by the $In_{0.53}Ga_{0.47}As$ bottom cell with no change in all cell parameters. By simulating $In_{0.53}Ga_{0.47}As$ cell, we can notice that it has an V_{OC} greater than that of Ge cell, since $In_{0.53}Ga_{0.47}As$ has higher energy bandgap than Ge cell. Furthermore, although the short circuit current of $In_{0.53}Ga_{0.47}As$ sub-cell is slightly higher than the short circuit current of Ge cell, it does not affect on the perfomance of the TJ solar cell as the current is limited by the top InGaP sub-cell.

$In_{0.53}Ga_{0.47}As$ has a lattice constant of 5.8687 Å while GaAs lattice constant is 5.653 Å. Because of this mismatch in lattice constant, atoms create chemical bonds in heterointerface by adjusting their positions. This creates strain and cause crystal dislocation and structure imperfection in depth [15]. This mismatch forms traps acting as recombination centers which will affect adversely the short circuit current produced by the cell. In InGaP/GaAs/InGaAs TJ solar cell, the lattice mismatch will appear between the middle cell and bottom cell. To overcome this mismatch between GaAs and $In_{0.53}Ga_{0.47}As$ lattice constants, two buffer layers with heavily doping concentration are used in the design. The buffer layer increases the excess minority carriers to the junction. Additionally, these layers drive out the photo-generated carries with minimum recombination losses. The simulation results of InGaP/GaAs/Ge cell and InGaP/GaAs/InGaAs cell show that the InGaP/GaAs/InGaAs cell has higher V_{OC} than the other, however both produce the same J_{SC}. This results in the efficency obtained by InGaP/GaAs/InGaAs TJ solar cell is higher than that obtained by InGaP/GaAs/Ge TJ solar cell. Figure 3 shows the I-V characteristics for InGaP/GaAs/Ge and InGaP/GaAs/InGaAs TJ solar cells. It is clear that the open circuit voltage of InGaP/GaAs/InGaAs TJ solar cell (2.9 V) is higher than that of InGaP/GaAs/Ge TJ solar cell (2.76 V) and the short circuit current is almost the same for both cells.

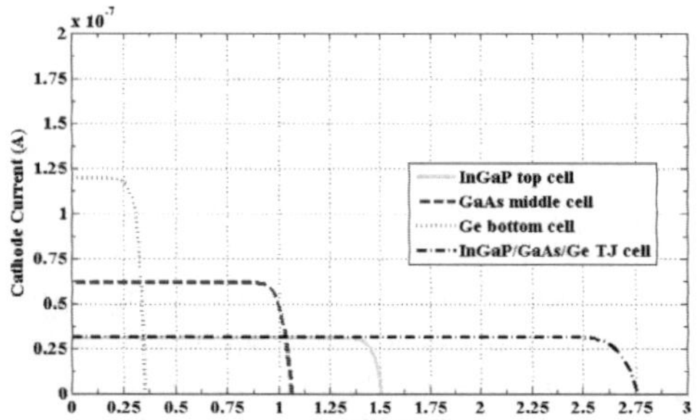

Figure 2. I-V characteristics for InGaP cell, GaAs cell, Ge cell and InGaP/GaAs/Ge TJ cell.

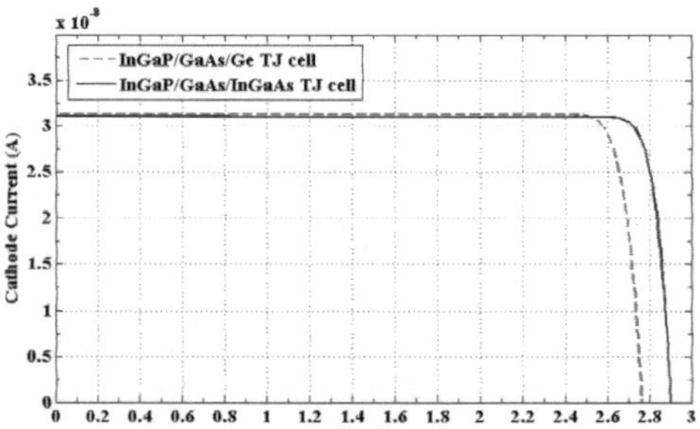

Figure 3. I-V characteristics of InGaP/GaAs/Ge and InGaP/GaAs/InGaAs TJ solar cells.

3. Optimization for InGaP/GaAs/InGaAs TJ solar cell

From the previous sections, we can conclude that the short circuit current is limited and controlled by both the top and middle cells. So, to optimize the short circuit current, the top and middle cells should be optimized. One way to optimize these cells is optimizing their thicknesses. In this section, a design of experiment (DOE) technique is accomplished in order to show how to optimize the thickness for each cell and hence the efficiency of the TJ solar cell. Table 1 summarizes the change in the performance parameters of solar cell with respect to the base thickness of the top InGaP cell and keeping all the cell parameters constant. It is clear that the optimized base thickess of the top cell is 600 nm at which the efficiency of TJ solar cell is 42.01%. After the top base thickness is optimized, the base of the middle cell could be varied in order to obtain the maximum allowable efficiency. Table 2 summarizes how the efficiency is varied using different values for the base thickness of the middle cell and keeping the base thickness of the top cell at 600 nm.

Table 1. Variation of the InGaP/GaAs/InGaAs TJ cell parameters w.r.t the top base thickness.

Thickness / Parameter	400 (nm)	500 (nm)	550 (nm)	600 (nm)	670 (nm)	820 (nm)
V_{OC} (V)	2.916	2.91	2.91	2.90	2.90	2.90
J_{SC} (mA/cm^2)	15.17	15.66	15.83	15.97	15.94	15.75
FF (%)	91.94	91.37	90.93	90.47	90.71	91.3
Efficiency (%)	40.64	41.63	41.86	42.01	41.99	41.69

Table 2. Variation of the InGaP/GaAs/InGaAs TJ cell parameters w.r.t the middle base thickness.

Thickness / Parameter	2 (μm)	2.4 (μm)	3 (μm)	3.4 (μm)	4 (μm)	5 (μm)
V_{OC} (V)	2.92	2.91	2.91	2.91	2.90	2.903
J_{SC} (mA/cm2)	14.57	15.01	15.83	15.76	15.97	15.977
FF (%)	92.16	91.76	90.93	90.77	90.47	91.09
Efficiency (%)	39.24	40.21	41.22	41.63	42.01	42.2

Although the highest possible efficiency could be obtained at a bottom base thickness of 5 μm, we choose the thickness of 4 μm because the device size will be increased (which results in a higher fabrication cost) with only an increase in the efficiency less than 0.2%.

Using the same previous methodology and geometric conditions of InGaP/GaAs/InGaAs TJ solar cell, an InGaP/GaAs/Ge TJ cell will be optimized. The results are as follows: V_{OC} = 2.767 V, J_{SC} = 16 mA/cm^2 and the η = 39.68%. The short circuit current of both cells are the same, so, the difference in efficiency results from the increase in open circuit voltage for InGaAs over Ge.

The generated P-V and the I-V characteristics of the optimized InGaP/GaAs/Ge TJ cell and the proposed InGaP/GaAs/InGaAs TJ cell are shown in figures 4a and 4b, respectively.

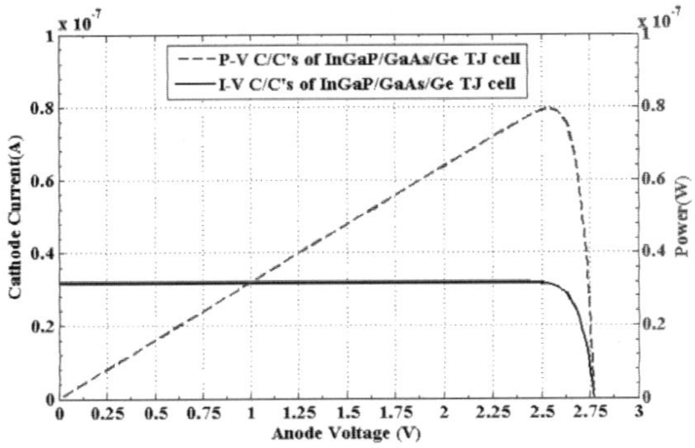

(a)

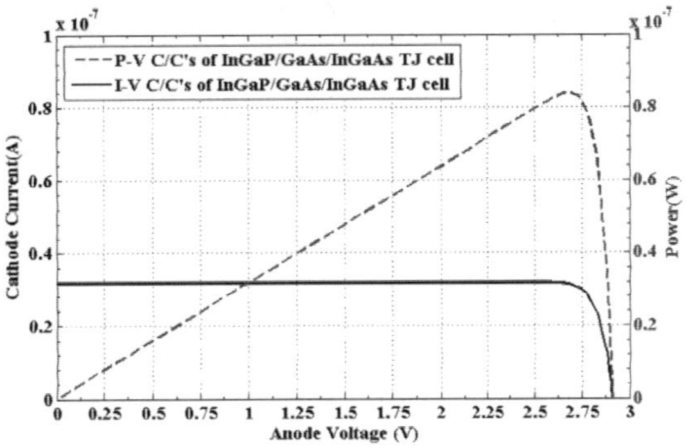

(b)

Figure 4. P-V and I-V characteristics for a) an optimized InGaP/GaAs/Ge TJ solar cell and b) a proposed InGaP/GaAs/InGaAs TJ cell device, under 1 sun of AM1.5G illumination.

From the P-V curves, we can notice that the maximum power obtained from the proposed InGaP/GaAs/InGaAs solar cell (42.259 mW/cm^2) is higher than that is obtained from the optimized

InGaP/GaAs/Ge solar cell (≈ 40 mW/cm^2). Also, for InGaP/GaAs/Ge TJ solar cell, the voltage at which maximum power is achieved is 2.675 V. Finally, the electrical characteristics for the proposed TJ solar cell are V_{OC} = 2.9 V, J_{SC} = 15.97 mA/cm^2 and the conversion efficiency η = 42.01%.

4. Results comparison

The obtained results from this optimized devices are now compared with corresponding experimental and theoritical published data [3, 7, 8, 16]. Table 3 shows a comparison between the proposed structure and the previously published data. Form table 3, it is clear that the performance of either optimized InGaP/GaAs/Ge TJ solar cell or the proposed InGaP/GaAs/InGaAs TJ solar cell is better than the other designs. As seen, the open circuit voltage as well as the short circuit cuurent are higher than that are introduced by [3, 7, 8, 16]. That leads to a higher efficiency for the proposed InGaP/GaAs/InGaAs cell or the optimized InGaP/GaAs/Ge cell ~~higher~~ than that obtained from the other published work.

Table 3. Comparison of proposed model with the different optimized TJ solar cell structures for spectrum AM 1.5G.

Solar Cell	Spectrum	Sun	V_{OC} (V)	J_{SC} (mA/cm^2)	*FF* (%)	η (%)
Huang *et al.* [16]	AM1.5	1	2.609	13.368	N/A	29.67
R.R. King *et al.* [7]	AM1.5	1	2.622	14.37	85	32
J.W. Leem *et al.* [3]	AM1.5	1	2.55	13.92	86.55	30.72
M.A. Green *et al.*[8]	AM1.5	1	2.66	13.1	85.6	34.5
Our optimized InGaP/GaAs/Ge cell	AM1.5	1	2.767	16	89.46	39.68
Our proposed InGaP/GaAs/InGaAs cell	AM1.5	1	2.9	15.97	90.47	42.01

5. Conclusion

In this work, MJSCs are designed and simulated using SILVACO TCAD tool. InGaP/GaAs/Ge and InGaP/GaAs/InGaAs TJ solar cells are simulated and the obtained results are compared. The efficiency of optimized InGaP/GaAs/Ge TJ solar cell is 39.61% while it is 42.01% for the proposed InGaP/GaAs/InGaAs cell. The increase in efficiency is due to increasing the open circuit voltage for the InGaAs sub-cell (0.48 V) over the Ge sub-cell (0.35 V) while the short circuit current still constant. All these results are obtained under 1 sun illumination of AM1.5 solar spectrum.

References:

[1] Leite M S, Woo R L, Munday J N, Hong W D, Mesropian S, Law D C and Atwater H A 2013 *Appl.Phys.Lett.* **102** 033901.
[2] Green M A, Emery K, Hishikawa Y, Warta W and Dunlop E D 2013 *Prog. Photovolt:Res.Appl.* **21** 1-11
[3] Jung Woo Leem, Jae Su Yu, Jong Nam Kim and Sam Kyu Noh 2014 *J. Korean Phys. Soc.* **64** 1561
[4] Babar M, Al-Ammar E A and Malik N H 2012 *Int J Energ Tech Pol.* **2** 44-53
[5] King R R *et al.* 2012 *Prog. Photovolt: Res. Appl.* **20** 801-815.
[6] Daniel C Law *et al.* 2013 In *Proceedings of SPIE - The International Society for Optical Engineering* **8876**
[7] Leite M S and Atwater H A 2012 *38th IEEE PVSC.* (Austin, TX, USA) pp. 002071-002075
[8] Martin A Green, Yoshihiro Hishikawa, Ewan D Dunlop, Dean H Levi, Jochen Hohl-Ebinger and Anita W.Y. Ho-Baillie 2018 *Prog. Photovolt: Res. Appl.* **26** 3-12
[9] Tsutagawa M H and Sherif Michael 2009 *34th IEEE PVSC.* (Philadelphia, USA) p 001954
[10] Nayak P P, Dutta J P and Mishra G P 2015 *Eng. Sci. Technol. an Int. J.* **18** 325–335.
[11] Silvaco Data Systems Inc, Silvaco Atlas user's manual, 2016.
[12] {https://refractiveindex.info/database/other/semiconductor alloys/GaP-InP/Schubert.yml}.

IOP Conf. Series: Materials Science and Engineering **446** (2018) 012010 doi:10.1088/1757-899X/446/1/012010

[13] {http://www.ioffe.ru/SVA/NSM/nk/index.html}.
[14] Sopra, (2014). {http://www.sopra-sa.com/indices.htm}.
[15] Panayiotis M 2002 "A novel approach for the development and optimization of state-of-the-art photovoltaic devices using Silvaco"
[16] Huang *et al.* 2011 *37th IEEE PVSC*. (Seattle, WA, USA) pp. 002071-002073

IOP Conf. Series: Materials Science and Engineering **446** (2018) 012011 doi:10.1088/1757-899X/446/1/012011

Effect of mixing of waste biomass in anaerobic digesters for production of biogas

Fosca Conti[1,2,*,a], Leonhard Wiedemann[1,b], Abdessamad Saidi[1,c], and Markus Goldbrunner[1,d]

[1]Institute of new Energy Systems, Technische Hochschule Ingolstadt (THI), Esplanade 10, 85049 Ingolstadt, Germany
[2]Department of Chemical Sciences, University of Padova, via Marzolo 1, 35141 Padova, Italy

Email: [a]fosca.conti@unipd.it (*corresponding author), [b]leonhard.wiedemann@thi.de, [c]abdessamad.saidi@thi.de, [d]markus.goldbrunner@thi.de

Abstract. Biogas technology is an important renewable bioenergy producer. The biogas generating process needs to be optimized to minimise the energy consumption due to the stirring of biomass slurry. Numerical simulations and laboratory experiments are economically and practically preferred over investigations of industrial scale biogas plants. Additionally, a strategic approach to model the reality in scientific laboratories is to use a rheological valid artificial chemical substrate to replace real biomass. The proposes of this study were (i) to investigate the mixing process in a 1:12 scaled-down home-made laboratory digester filled with a 0.3 wt% water-cellulose solution, (ii) to simulate the mixing process in the laboratory-scale digester using a computational fluid dynamics model, (iii) to validate the model by comparison of the simulation with laboratory experiments results obtained on the laboratory digester. Optical and acoustic measurements on the flow velocity inside the digester during the mixing process of the water-cellulose solutions indicate that the model based on computational fluid dynamics is valid. The data are presented and discussed in the paper.

1. Introduction

Human global energy consumption requires competitive energy forms and continual development of strategies concerning renewable energies. Sunlight, wind, waves and tides, geothermal phenomena and bioresources are common renewable energy sources. Bioresources, commonly known as biomass, produces bioenergy in form of heat, electricity, biofuels and has gained significant interest as an alternative to petroleum fuels, especially in the last thirty years [1].

In comparison to wind and solar power, which are fluctuating energy suppliers, the biomass power has the ability to balance these fluctuating energy producers due to its capacity to produce energy continuously and to realize demand oriented production. Different kinds of processes have been developed to extract energy from biomass, depending on the chemical and physical state of the feedstock itself and, of course, the final use and destination [2]. The most common type of bioenergy is combustion of lignocellulose pre-treated biomass (e.g. wood pellet), oil crops or animal fat. Not only are these sources suitable for heating, but they can also be combined with electricity production. For sludge, liquid and wet organic materials, anaerobic digestion is currently the best-suited option for producing electricity and/or heat from biomass, although its economic viability relies heavily on the availability of low cost feedstock. Gasification is also a good renewable alternative for heat and power generation. It consists of the treatment of organic matter at high temperatures without combustion, producing CO, CO_2 and H_2 gas. Biofuels are instead used for the transport sector: bioethanol from starch and sugar crops and biodiesel from oil crops and residual oils and fats, mostly.

Anaerobic digestion systems use waste biomass to produce renewable energy in the form of biogas. To maximize gas production, optimal conditions for decomposition of organic materials in digesters are necessary, such as mixing to homogenize the biomass volume. Cost-benefit analysis indicate that mixing is the highest contributor to the total energy consumption in biogas plants. To reduce this parasitic contributor testing of different mixing regimes is required, i. a. spatial and operational conditions of agitators [3-5]. Since it is not economically feasible to perform these tests on an industrial scale biogas plant, laboratory scale experiments emerge as a convenient and appropriate approach to study the mixing in full-scale anaerobic digesters. Recently we have designed a scaled-down digester [6, 7] and select an artificial substrate [8] with the aim to mimic in our scientific laboratory what is the mixing process in a real fermenter filled with biomass. In this paper, we investigate the mixing process using a 1:12 scaled down laboratory digester, an artificial chemical substrate based on cellulose, optical and acoustic spectroscopies and computational fluid dynamics (CFD) simulations. Indeed, computational simulations using CFD is a powerful tool to assess mixing quality of different system configurations and several model are described in the literature [9-12]. In this paper we use the well-known Ostwald-de Waele power law model for non-Newtonian fluids [13, 14]. Technical challenges and difficulties of mixing configurations can be rationalized in the model to obtain in valuable interval of time important parameters to evaluate the efficiency of the process. The next phase is to scale up the resulting parameters to correlate the dynamic in the laboratory digester to that in the real fermenter of the industrial scale biogas plants.

2. Materials and methods

2. 1. Preparation of artificial substrate

The artificial chemical substrate was prepared using a sodium carboxy-methyl cellulose purchased from Dow Wolff Cellulosics GmbH, Germany, with the commercial name of Walocel CRT 30000TM. Aqueous Walocel solution with 0.3 wt% dry matter content was prepared dissolving 2.5 Kg of powder in 800 l of water at 20 °C to obtain rheological characteristics comparable to the biomass slurry [8]. The artificial substrate was mixed intermittently for three days to obtain a homogeneous solution.

2.2 Experimental laboratory digester setup

A 1:12 scaled-down laboratory digester was designed and constructed in polymethylmethacrylat to replicate the digester tank of full-scale biogas plants. Mixing was achieved by two IKA Eurostar 100 control overhead stirrers placed opposite in the tank. The motors are equipped with variable speed drivers for selecting and controlling motor speed. Six access holes on the tank allow to insert the two stirring shafts at different heights (levels) and angles. Three types of agitator were installed: a paddle equipped with four rectangular plates and two propellers with three wings and characteristic diameter of 7.5 cm and 12.5 cm (see Figure 1).

2.3 Setup for optical and acoustic investigation

The mixing process has been investigated in the laboratory digester in terms of flow velocity and flow pattern inside the tank. Optical spectroscopy (Particle Image Velocimetry, PIV) and acoustic techniques (Acoustic Doppler Velocimetry, ADV) have been used.

For PIV measurements, seeding particles (microporous polyamide with average particle diameter of 20 μm) were injected in the water-cellulose fluid and illuminated with a Nd-YAG laser sheet. The trajectories of the particles were captured by a high speed camera in consecutive images. By cross-correlating the images, the fluid velocity was calculated to obtain velocity vector maps, statistics and spatial correlations of the artificial substrate.

For ADV measurements, the water-cellulose fluid velocity was directly measured using a Vectrino Velocimeter operating with the Doppler effect principle and a bistatic sonar, i.e. the ADV velocimeter uses separate transmit and receiver beams, in contrast to standard Doppler profiles. The reflection (echo) from the particles suspended in the water was picked up by four receive transducers.

2.4 Computational fluid dynamics (CFD) analysis
To simulate the mixing process, the commercial computational fluid dynamics (CFD) software StarCCM+ from CD-Adapco Siemens was used. Simulation aspects cover geometry design of the tank, grid generation, selection of physical models and solvers, data evaluation and optimization.

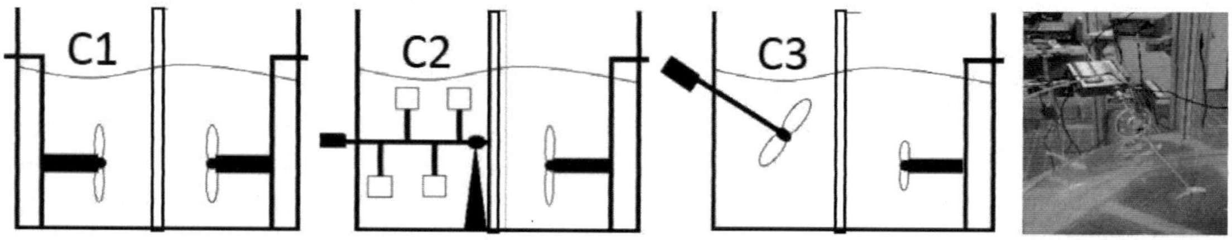

Figure 1. Scheme of the three mixing configurations, which are used in the laboratory digester tank to validate the CFD simulations. C1: two submersed big propellers produce a symmetric flow. C2: one big propeller and one paddle. C3: one small propeller and one incline big propeller. Foto of the incline big propeller mixer used in C3.

3. Results and discussion

3.1 Effect of mixing configurations
To study the mixing process, three combinations of propeller and paddle were configured in the laboratory tank (Figure 1). In general, the efficiency of the mixing process inside a digester depends on a large number of parameters, which are related to the substrate rheology, design of the tank (geometry and size) and stirrers (quantity, locations, direction), and operational procedure (continuous or intermitted substrate feeding, bulk liquid temperature, mixing velocity, and continuous or intermitted mixing). The three configurations, C1-C3, were selected on the basis of four factors: (a) the intention to not be limited to a single mixer but to use different types of mixers, which represent the state of the art in biogas plants, (b) the idea to use configurations comparable to full-scale applications to obtain results which may be valuable to the operational procedures in biogas plants (c) the plan to start with a simple flow dynamic (in C1) and to increase its complexity (in C2 and in C3), and (d) the concept to have symmetrical (C1) and asymmetrical (C2 and C3) flows inside the laboratory digester.

3.2 Effect of flow velocity
The distribution of the homogeneous substrate was evaluated with PIV and ADV measurements on the flow velocity at three levels (height) in the tank with 10 locations each (marked from 1 to 5 and from 6 to 10 in the CFD scenario of Figure 2 and 3). The y components of the flow velocity of the water-cellulose mixture in the ten positions inside the laboratory digester are plotted in Figure 2. The results refer to the mixing configuration C1 of Figure 1.

The results indicate a development of the flow velocity from the wall of the digester to the center. Towards the center, the slower moving fluid is accompanied by less mixing dynamics. Lower fluid velocities were apparent near the center of the digester. This suggested lower quality of the mixing within the region, since slow moving zones represent less mixed zones. The presence of a dead zone in the center is merely related to the movement of the flow in the plane parallel to the bottom of the digester (xy). In the z axis the flow appears slightly eddying and with up and down movement. The velocity values in Figure 2 ranged in the interval 5 - 10 cm/s. The velocity depends on the friction at the wall, then increases to a maximum of ca. 0.1 m/s, and then nearly no flow is present in the central regions.

For the C1 configuration, a symmetry in the flow dynamics was expected between the 1-5 and 6-10 regions, such that the results in positions 1 and 6 should be comparable and also those of positions 5 and 10. This expectation was validated by the PIV and ADV experimental data, which confirm the precision of the used PIV and ADV setups and the accuracy of the experimental procedure. The similarity of PIV and ADV data suggest the possibility of limiting the investigation to only ADV technology. Indeed, the accuracy of the ADV measurements (error of the ±0.5% of the value) is much higher due to the high error of the PIV measurements (±5% of the value).

3.3 CFD evaluation

One main purpose of the present research was to obtain experimental results which could be used to validate a CFD model recently developed. The parametric settings are listed in Table 1.

Table 1. CFD parametric settings for the study of mixing configurations C1-C3.

	parameter	specification		parameter	specification
1	Type of mixer	Submersed 3-bladed propeller mixer	**6**	Height	Positioning up/down of mixer
2	Number of mixer		**7**	Placement in/out	Positioning at different radii
3	distribution of mixers	Placement of mixers to one another	**8**	Elevation angle up/down	Angle mixer orientation vertical direction
4	Digester geometry	Diameter of digester and height of fluid level filled	**9**	Mode of operation	Continuous/time-sequence
5	Side angle	Angle mixer orientation horizontal (sidewards)	**10**	Rotational speed	Mixer rotational speed [rpm]
			11	Rheology	

For computer simulations with StarCCM+, the Non-Newtonian generalized power law was used for the fluid rheology. A consistency factor of 0.05 [Pa•s^n] and flow index of 0.25 were specified. In the CFD calculation, the k-ε turbulence model was used, integrated with steady-state and moving reference frame model (MRF) to implement rotation and static regions.

Data obtained from the PIV and ADV experimental measurements were compared with the CFD calculated flow velocities. In Figure 2 are plotted the results obtained for the y component of the flow velocity in configuration C1.

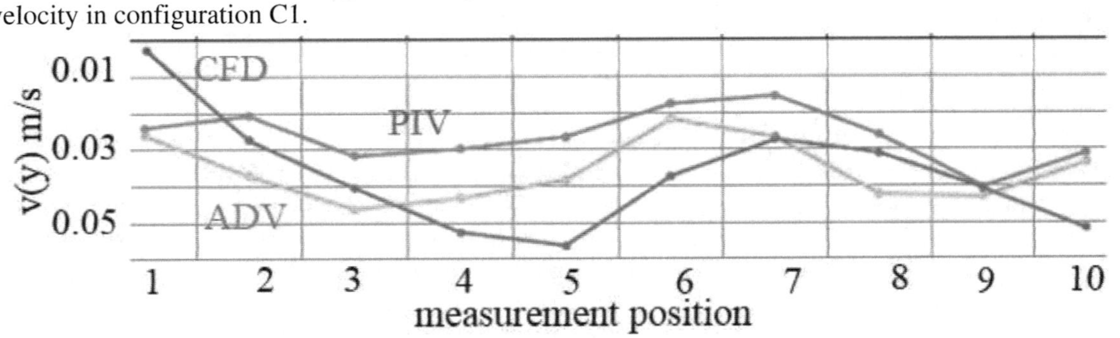

Figure 2. Flow y-velocity results obtained with PIV, ADV, and CFD investigations. Measuring positions 1-10 refer to the ten locations in the digester depicted in Figure 3.

The correlation between the experimental data obtained with PIV and ADV technologies and the theoretical data obtained with the CFD simulation is very promising. CFD simulation is advantageous over laboratory experiments due to the drastically reduced timescale required to visualize flows. Experiments on the laboratory digester were time consuming and limited by the number of configurations and parameters which could be checked. For example, the investigation of the flow dynamic in the 50-250 rpm range at steps of 50 rpm needed weeks of measurements but only hours of computer simulation. The discrepancy in time required is also evident when changing the angle of the shaft of the stirrer from 30° to 60°. However, CFD simulations alone, without any confirmation from experimental results, cannot be justified. Thus, a combination of the two methodologies must be applied.

3.4 Effect of torque moment

Additional measurement of the torque moment on the stirrer-blades indicate the power consumption for each configuration. Values in the range of 0.0045Nm-0.0065Nm (measurement error 0.1% of full scale 1 Nm) were found. The results indicate that the use of the smaller propeller required more energy and produced less hydraulic torque. The consideration that a large propeller is less energy demanding is useful for mixing design and selection of stirring configurations in real digesters.

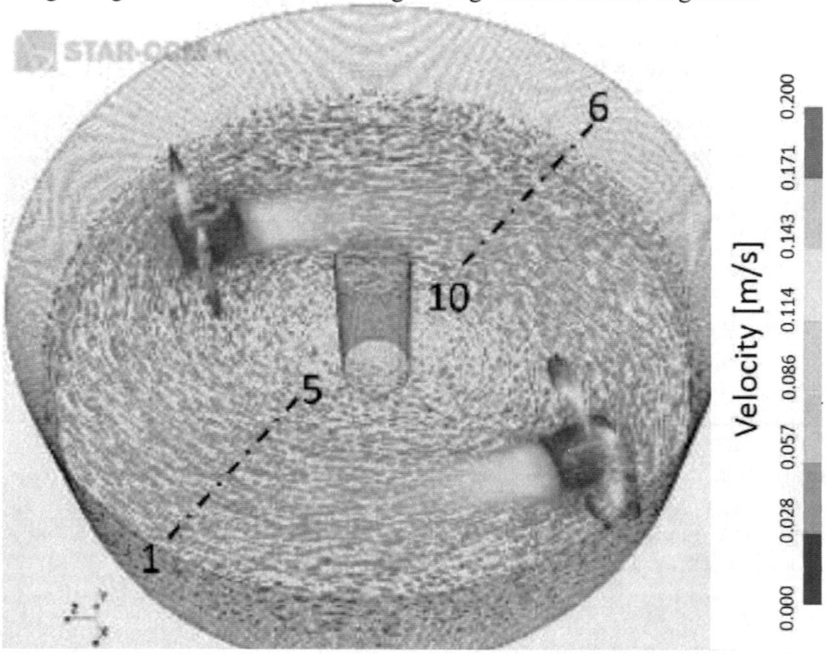

Figure 3. Top view scenario of the flow velocity vector obtained using the CFD model. Positions 1-10 inside the digester refer to the ten measuring locations investigated by PIV and ADV (Figure 2).

4. Conclusions

A scaled down laboratory digester equipped with mixing systems, and optical and acoustic setups was designed and successfully constructed to investigate the mixing behavior of an artificial chemical substrate, used to substitute real biomass. In this paper PIV and ADV laboratory results are shown to determine flow characteristics of sludge in digesters. Results on power consumption, torque moment and flow velocity confirmed the accuracy of the experimental procedure. The data are comparable to CFD numerical calculations so that the laboratory experiments can be used as validation tool to the CFD simulating model. The approach allows at a much lower cost than laboratory experiments or investigations at full-scale.

In real case scenario, the condition will be anaerobic, temperature substrate quality will be different, the microorganisms will be present, etc. These points limit the applicability of any result and model.

On the other hand, researches on industrial scale biogas plant are not economically feasible. There is a need for deeper insight on the complexity of this fascinating biological process. The presented study suggests one possible successful strategy. The next step is studying the transferability of the results obtained with PIV, ADV and CFD on the scaled down laboratory digester to the industrial scale biogas plants. A full examination of this scaling up process is in plan to be conducted in future studies.

References

[1] Sikarwar V S, Zhao M, Clough P, Yao J, Zhong X, Memon M Z, Shah N, Anthony E J and Fennell P S 2016 *Energy Environ Sci.* **9** 2939.
[2] Mao C, Feng Y, Wang X and Ren G 2015 *Renewable Sustainable Energy Rev.* **45** 540.
[3] Wiedemann L, Conti F, Janus T, Sonnleitner M, Zoerner W and Goldbrunner M 2017 *Chem. Eng. Technol.* **40** 238.
[4] Conti F, Wiedemann L, Janus T, Sonnleitner M, Zörner W and Goldbrunner M 2016 *Proc. 10. Rostock Bioenergy Forum* (Rostock) ISBN: 978-3-86009-433-4, 58, 445-450, 2016.
[5] Wiedemann L, Conti F, Sonnleitner M, Saidi A and Goldbrunner M 2017 *Proc. 25th European Biomass Conf. and Exhibition EUBCE* (Stockholm) p 889-892, ISBN: 978-88-89407-17-2, DOI: 10.5071/25thEUBCE2017-2CV.4.14.
[6] Wiedemann L, Conti F, Janus T, Sonnleitner M, Zoerner W and Goldbrunner M 2017 *Proc. Int. Conf. Progress in Biogas IV* (Stuttgart) ISBN: 978-3-940706-09-6, p 112.
[7] Wiedemann L, Conti F, Janus T, Sonnleitner M, Zoerner W and Goldbrunner M 2017 *Proc. Int. Conf. Progress in Biogas IV* (Stuttgart) ISBN: 978-3-940706-09-6, p 147.
[8] Conti F, Wiedemann L, Sonnleitner M and Goldbrunner M 2018 *New J. Chem.* **42** 1099.
[9] Dapelo D, Alberini F, Bridgeman J 2015 *Water Res.* **85** 597
[10] Lomtscher A, Jobst K, Fogel S, Rostalski K, Stempin S, Kraume M 2017 *Flow Meas. Instrum.* **53** 56
[11] Wu B 2011 *Water Res.* **45** 2082.
[12] Yan X, Wu Q, Sun J, Liang P, Zhang X, Xiao K, Huuang X 2016 *Bioresourc. Technol.* **200** 328.
[13] Natarajan VP and Suppes GJ 1997 *Fuel.* **76** 1527.
[14] Macosko CW 1994 *Rheology Principles, Measurements, and Applications*, VCH, New York.

Acknowledgement

The work was financially supported by the German Federal Ministry of Education and Research within the program “Forschung an Fachhochschulen” under the grant “BioOpt-Mix 03FH031PX4”. FC is grateful to the University of Padova for providing the opportunity to spend research periods at the Technische Hochschule Ingolstadt.

IOP Conf. Series: Materials Science and Engineering **446** (2018) 012012 doi:10.1088/1757-899X/446/1/012012

Membrane separation process for small scaled partial biogas upgrading

Abdessamad Saidi[1,a], Fosca Conti[1,2,*,b], Matthias Sonnleitner[1,c] and Markus Goldbrunner[1,d]

[1]Institute of new Energy Systems, Technische Hochschule Ingolstadt, Esplanade 10, 85049 Ingolstadt, Germany
[2]Department of Chemical Sciences, University of Padova, via Marzolo 1, 35141 Padova, Italy

Email: [a]abdessamad.saidi@thi.de, [b]fosca.conti@unipd.it (*corresponding author), [c]matthias.sonnleitner@thi.de, [d]markus.goldbrunner@thi.de

Abstract. Biogas upgrading is actually limited to the production of biomethane as natural gas substitute. To realize the according gas quality a cost intensive methane enrichment is required, wherefore biogas upgrading is actually merely cost-efficient in case of high production rates. Since the energy and cost effort disproportionately increases with the required product gas purity partial biogas upgrading for decentralized utilization represents a promising utilization approach for farm based applications. Among the available technologies for CO_2-separation the gas permeable membrane has high potential for small scaled biogas upgrading. Within the present study a model based analysis to determine the savings potential of a membrane based upgrading system is performed.

1. Introduction

To achieve climate protection goals at national and international levels, the expansion of renewable sustainable energies has been quantified with specified obligatory targets initialized by the Kyoto decisions concerning the reduction of greenhouse gases (GHG). In Germany, a GHG emissions reduction by at least 40 % until 2020 as compared to 1990 and by 80 – 90 % until 2050 in comparison to 1990 is persued [1].

Besides energy efficiency measures, the expansion of energy production from renewable sources is a mainstay to achieve these ambitious goals. The Directive for renewable energy sources of the European Union (2009/28/EG) determines a share of renewable energies of 18 % of the gross energy consumption until 2020 [2].

In Germany the implementation of renewable energy into the structure of energy supply has been expanding continuously within the recent years. In 2016, a total feed-in power of 188.3 TWh_{el} is occurred which implies a share of 31.7 % of the electricity consumption [3]. With 32.37 TWh_{el}, biogas provides about 17.2 % of electricity generation from renewable energy sources and therefore makes a significant contribution to sustainable electricity production besides wind energy and photovoltaics [3].

The massive build-up of biogas plants over the last decade was stimulated by governmental funding and legal framework of the Renewable Energy Sources Act. To preserve the existing plants after the financial funding has expired the economic efficiency of biogas processes has to be increased and new revenue opportunities have to be uncovered.

As an alternative utilization pathway to conventional CHP conversion for electricity and heat production biogas upgrading has been increasingly applied in the recent years. The product gas, so called biomethane, can be injected into the natural gas grid. To meet the high quality standards for the use as natural gas substitute various gas treatment steps are required. Mainly biogas upgrading includes gas drying, desulphurization, removal of further harmful trace components and CO_2-separation to improve the calorific value [4].

Due to high investment and operating costs, this option is constrained to high production rates. Since the CO_2 separation represents the most energy and cost intensive sub process, a reduced methane enrichment enables profitable farm based biogas upgrading on a small scale.

Within the present research project, the approach of decentralized utilization of partially upgraded biogas as fuel for agricultural machinery is investigated. Experimental studies on a reequipped single-cylinder diesel revealed, that performance equivalent engine operation can be realized with a CO_2-content up to 25 %. This determined concentration represents the superordinate goal for the design of the upgrading system [5].

To remove the CO_2 fraction from gas mixtures various methods have been developed and optimized in the gas processing industry within the recent decades. For biogas upgrading the following processes are available [4]:

Table 1. Available biogas upgrading technologies [4]

	CO_2 separation processes
Solvent demand	Water scrubbing Organic solvents Amine scrubbing
No solvent demand	Pressure swing Cryogenic separation Membrane separation

Due to its low complexity and equipment demand the membrane separation shows high potential for small scaled applications among the currently available upgrading technologies and therefore represents a promising option for the present research issue.

The separation mechanism relies on the preferential transfer of one gas fraction from a mixture through a semi-permeable membrane material. Membranes can be grouped into high pressure membranes which have gases present on each side of the membrane, and low pressure systems, which have a liquid adsorbent on one side of the membrane wall. High pressure membranes are widely used for biogas upgrading. In a membrane system, high concentrations of contaminants such as H_2S and moisture are generally reduced prior to methane enrichment. High pressure membrane separation is typically undertaken at 8–15 bar [6]. Biogas is generally upgraded in a multi stage process to yield a final CH_4 concentration of 96 %. Waste gases from the preliminary stages are recycled within the process to enhance CH_4 capture whilst waste gas from the final stage, which maycontain10–20% CH_4, is either flared, used for heat production [6] or captured catalytically. This technology has been applied for some time for the upgrading of natural gas.

2. Methods

2.1 Upgrading concept

The approach of reduced CO_2-separation provides the option for further simplifications of the conventional membrane separation process (e.g. Figure ***1***).

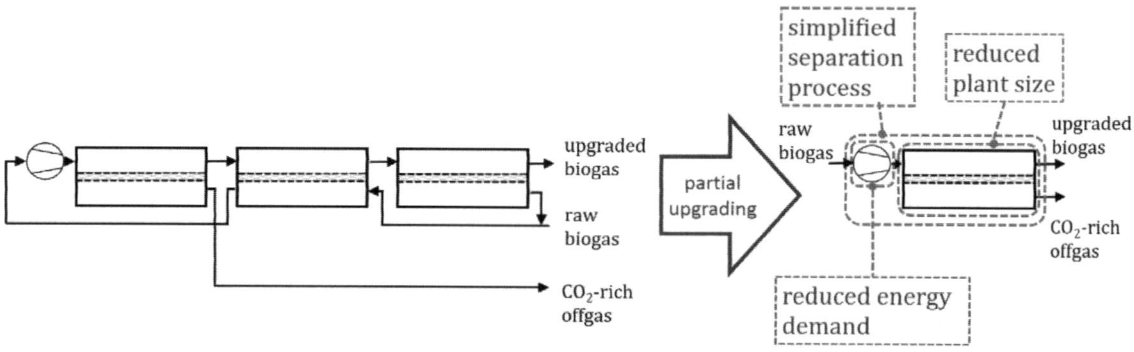

Figure 1. Simplification options for membrane separation process

The most significant process simplification is given by the reduction of the separation stages. In principle, the number of stages does not represent a limiting factor for the product gas purity, since any composition of the product gas can be realized by a single membrane unit. Rather, the multi-stage process provides an option to mitigate the tradeoff between the product gas purity and mass efficiency. Since the methane rich permeate of the final stages is recirculated by the feed flow, the multi stage configuration enables a significant reduction of the methane losses.

Regarding overall process efficiency and after treatment of the waste gas the methane losses define an essential criterion for conventional upgrading systems. However, for small scaled biogas upgrading, the methane content of the permeate flow is of subordinate significance. Taking typical biogas production rates into account, the comparatively low gas flow provided as fuel for agricultural machinery allows the recirculation of the waste gas stream without a significant impact on the raw biogas composition. The impact of the recirculation for different gas flow rates and methane slip values (share of separated methane) can be determined by simple mass balance calculations and is illustrated in Figure **2**.

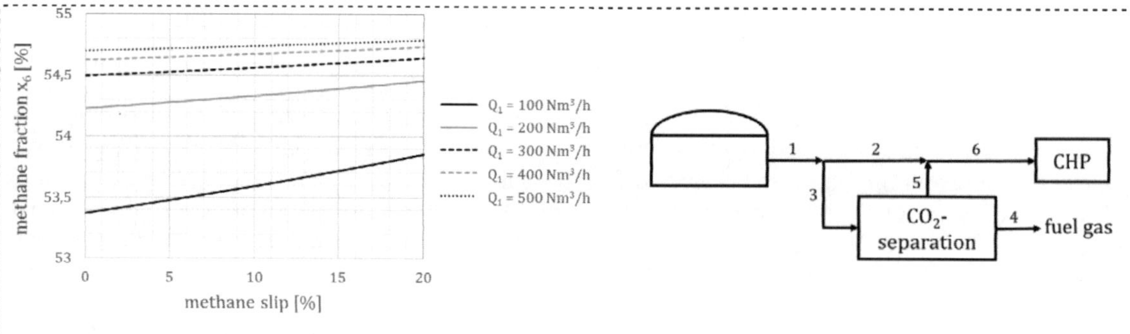

Figure 2. Impact of the methane slip on the raw biogas composition with recirculation of the offgas

As depicted, the raw biogas plus recirculated gas flow show decreased reduction of the methane content with increased methane slip. However, since the impact on the gas composition is negligibly small, a recirculation of the permeate gas can be performed.

The product gas purity and separation efficiency of the membrane process are strongly related to the operating pressure. The low purity requirements afford the option to reduce the energy demand of the separation process by lowering the compressor capacity.

A disadvantage of the simplified configuration is the reduced degrees of freedom with fewer adjustable process parameters. While the tradeoff between the product gas purity, the operation pressure, and product gas flow rate can be mitigated in multi stage separation with process design and off gas recirculation, a decoupling of these parameters is not possible in a single membrane unit. This issue is illustrated in detail based on the following explanations.

2.2 Model equations

The permeable separation of gas components represents an established gas processing method in different industry sectors, giving rise to the development of various modelling approaches in recent decades.

Table 2. Existing modelling options for membrane separation

Developer	Year	Reference
Coker et al.	1999	[7]
Marriott et al.	2001	[8]
Kalids et al.	2000	[9]
Giglia et al.	1991	[10]
Davis	2002	[11]
Katoh et al.	2010	[12]
Makaruk et al.	2009	[13]
Mota et al.	2002	[14]
Rautenbach et al.	1996	[15]
Scholz et al.	2012	[16]
Ohlrogge et al.	2005	[17]

Due to the high thematic consistency and appropriate compromise between accuracy and computing time, the model calculations of Makaruk et al. [13] are chosen among these methods for the present research objective. The approach provides a discretization of the continuous separation process. For one component i, the discretization is based on the finite difference method in a domain including the number c of discrete points (Figure 13).

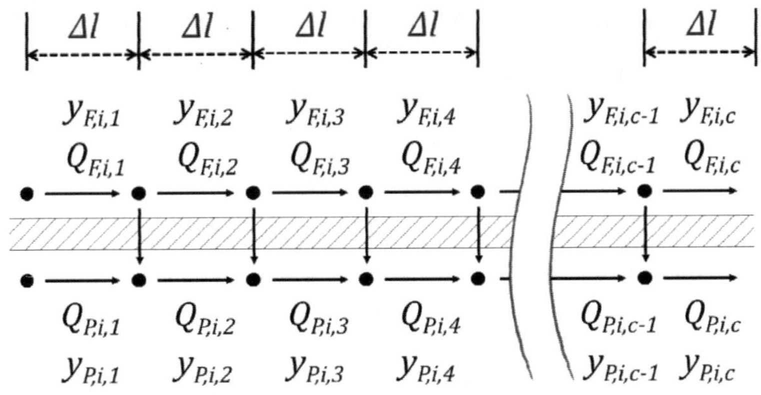

Figure 3. Discretization of the continuous membrane separation process

The corresponding conservation equations of the feed and permeate stream for a co-current separation process are presented as follows [13]:

$$Q_{F,i,j}^{n+(1/2)} = Q_{F,i,j-1}^{n} - \Delta l \Pi_{i,j}^{n} \left(y_{F,i,j}^{n} p_{F,j}^{n} - y_{P,i,j}^{n} p_{P,j}^{n} \right) s\pi d \qquad (1)$$

$$Q_{P,i,j}^{n+(1/2)} = Q_{P,i,j-1}^{n} + \Delta l \Pi_{i,j}^{n} \left(y_{F,i,j}^{n} p_{F,j}^{n} - y_{P,i,j}^{n} p_{P,j}^{n} \right) s \pi d \tag{2}$$

$$Q_{sep,i,j}^{n+(1/2)} = \Delta l \Pi_{i,j}^{n} \left(y_{F,i,j}^{n} p_{F,j}^{n} - y_{P,i,j}^{n} p_{P,j}^{n} \right) s \pi d \tag{3}$$

The state variables within the iterative solver are characterized by the index n. Since the separation efficiency is strongly related to the operating pressure, an accurate determination of the pressure gradient along the membrane unit is essential to generate a valid separation model. Taking the flow characteristic of membrane processes into account, the dissipative pressure loss has to be considered in addition to the pressure decrease caused by permeation. Under the assumption of a laminar flow, the pressure loss can be calculated via the Hagen-Poiseulle law. The corresponding equation for a discretized system is shown as follows [18].

$$\Delta p_F = \frac{\sum_{i=1}^{k} Q_{F,i} \mu_G 128 \Delta l p_{STP} T_F}{s \pi D^4 p_F T_{STP}} \tag{4}$$

Besides the separation pressure, which represents a process variable, the permeability as a material property has to be known. While several reported values for membrane materials are available in literature, membrane permeability of commercially available membranes are often concealed. To consider the actual state of the art, the present model is based on simulations provided by a co-developer of the SEPURAN® Green membrane (Evonik industries), which ranks among the most applied products for permeable biogas upgrading in Germany. Since the provided simulations are derived from real operations performed by pilot and commercial plants, validation of the present model is fulfilled to a great extent.

3. Results and discussion

A sensitivity analysis was conducted on the simulation with varying gas flow rate and operating pressure. The gas composition on the permeate and retentate side along the membrane for a feed flow of 10 and 11 m^3_{STP}/h is shown in Figure **4**.

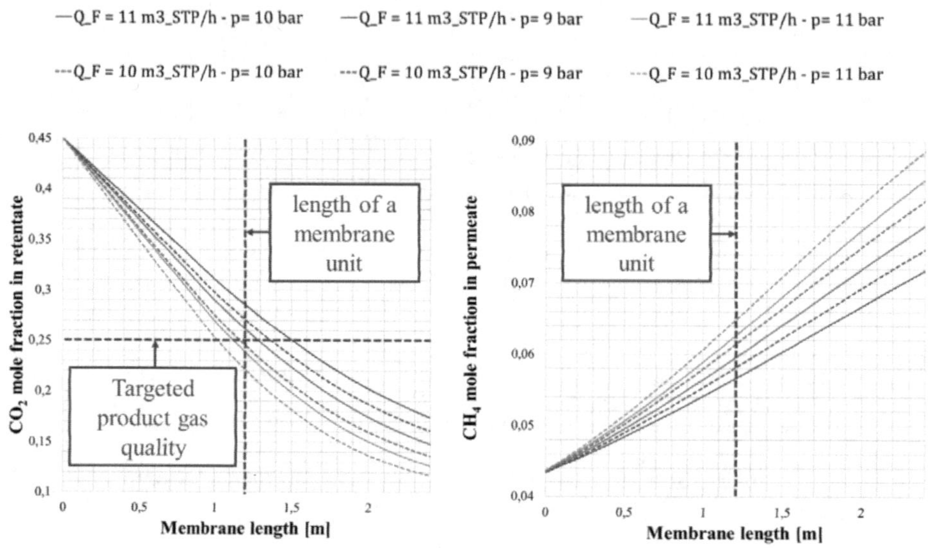

Figure 4. Gas composition on the retentate (left) and permeate side (right) of the membrane

As expected, the permeation of the CO_2 fraction is advanced by pressure increase, leading to lower CO_2-concentrations of the retentate flow (e.g. Figure **4** left). For higher volume flows, an appropriate rise of the operation pressure is required to enable adequate product gas purity. Concomitantly, a pressure increase is also connected to higher methane concentration in the permeate flow (e.g. Figure **4** right), implying lower methane recovery. According to the explanations around Figure **2**, the methane content on the retentate side should not be considered as methane losses. Nevertheless, in terms of conservation of mass, an increased methane permeation requires a higher feed demand to reach a targeted product gas amount and composition.

A summarizing illustration of the product gas composition over the entire variation range for flow rate and operation pressure is shown in Figure 5.

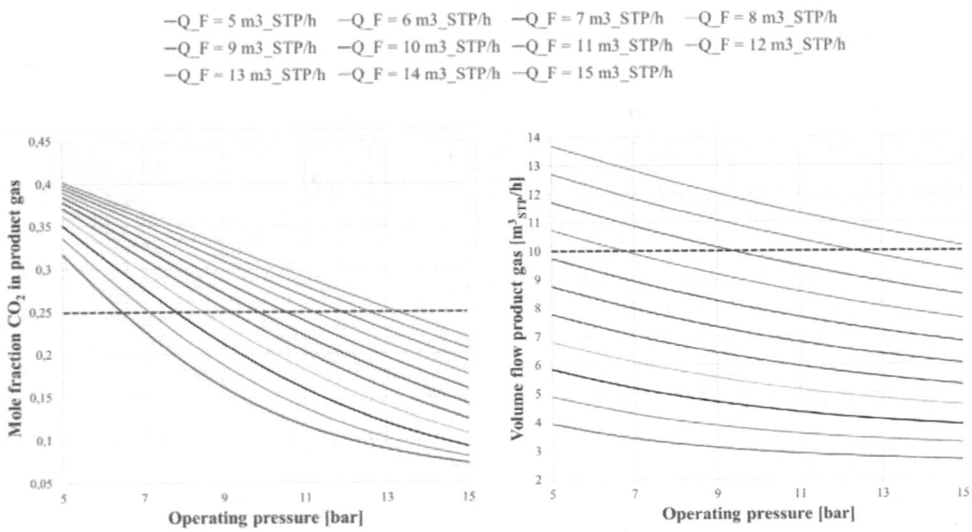

Figure 5. Product gas composition (left) and volume flow (right) depending on the operating pressure

For a pre-specified raw biogas flow rate, the operating pressure is the single lever to achieve a targeted gas composition. For example, a gas flow with a CO_2 content of 25 % and a flow rate of 10 m^3_{STP}/h can be realized with an operation pressure of about 12.4 bar and a raw biogas flow of approximately 14 m^3_{STP}/h. Decoupling of the process parameters is not possible. A single stage membrane separation is severely limited by the degrees of freedom to accommodate fluctuating raw biogas compositions.

4. Conclusions

The membrane separation process shows high potential as economic methane enrichment technology for small scaled applications. The approach of a reduced CO_2-separation provides the option for further simplifications of the conventional process. Furthermore, the idea enables lower operating pressures and therefore contributes to an increased cost-efficiency. Since the CO_2-separation merely represents the key step of the upgrading system, an overall model based investigation including the further gas processing steps as well as an economic efficiency calculation has to be performed to determine the total savings potential.

References

[1] *Energiekonzept für eine umweltschonende, zuverlässige und bezahlbare Energieversorgung,*

Federal Ministry for Economic Affairs and Energy, Berlin 2010.

[2] *Directive 2009/28/EC*, European Parliament and Council, Brussels 2009.

[3] *Erneuerbare Energien in Deutschland*, Umweltbundesamt, Dessau-Roßlau 2017.

[4] Poeschl, M., Ward, S., Owende, P. 2010. Prospects *for expanded utilization of biogas in Germany*, Ren. and Sust. En. Rev. 14, 1782–1797 [1] Davalos J. et. al. 1976 *J. Chem. Eng. Data* **21** 1.

[5] Saidi, A., Beringer, M., Sonnleitner, M., Männl, U., Innerhofer, S., Goldbrunner, M., *Combustion and exhaust emission characteristics of a dual fuel engine operated with low caloric biogas,* Poster: Progress in Biogas IV, 8-11/03/2017, Stuttgart.

[6] Wellinger, A., Lindberg, A., *Biogas Upgrading and Utilisation*. IEA Bioenergy Task 24, 2005.

[7] Coker et al., *Nonisothermal model for gas separation*, AIChE Journal, p. 1451–1468, 1999.

[8] Marriott et al., *Detailed mathematical modelling of membrane*, Computers & Chemical Engineering, p. 693–700, 2001.

[9] Kalids et al. , *Simulation of multicomponent gas separation in a hollow fiber membrane by orthogonal collocation–hydrogen recovery from refinery gases*, Journal of Membrane Science, p. 61–71, 2000.

[10] Giglia et al. , *Mathematical and experimental analysis of gas separation by hollow fiber membranes, Industrial & Engineering Chemistry*, p. 1239–1248, 1991.

[11] Davis et. al. , Simple gas permeation and pervaporation membrane unit operation models for process simulators, Chemical Engineering & Technology, p. 717–722, 2002.

[12] Katoh et al. , *Dynamic simulation of multicomponent gas separation by hollow-fiber membrane module: Nonideal mixing flows in permeate and residue sides using the tanks-in-series model.*Separation and Purification Technology, 2010.

[13] Makaruk et al. , *Numerical algorithm for modelling multicomponent multipermeator systems, " Journal of Membrane Science*, p. 258–265, 2009.

[14] Mota et al. , *Simulation of a new hybrid membrane/pressure swing adsorption process for gas separation, Desalination*, p. 275–280, 2002.

[15] Rautenbach et al. , *Simulation and design of membrane plants with Aspen Plus*, Chemical Engineering & Technology, p. 391–397, 1996.

[16] Scholz et al. et al., *Modeling gas permeation by linking non-ideal effects*, Industrial & Engineering Chemistry Research, 2012.

[17] Ohlrogge et al. , *Membranverfahren zur Abtrennung organischer Dämpfe in der chemischen und petrochemischen Industrie*, Chemie Ingenieur Technik, p. 527–537, 2005.

[18] James P. Hartnett; Milivoje K., *Heat Transfer to Newtonian and Non-Newtonian Fluids in* Rectangular *Ducts*. Advances in Heat Transfer, Volume 19, 1989.

Acknowledgement

The work was financially supported by the German Federal Ministry of Education and Research within the program "Forschung an Fachhochschulen" under the grant "22402215". FC is grateful to the University of Padova, Italy, for providing the opportunity to spend research periods at the Technische Hochschule Ingolstadt.

IOP Conf. Series: Materials Science and Engineering **446** (2018) 012013 doi:10.1088/1757-899X/446/1/012013

Ethylene glycol dry reforming on Ni/Al2O3 catalyst for syngas generation

M B Bahari[1], L N Jun[1], M-N N Shafiqah[1], F Fayaz[1] and D-V N Vo[1,2,*]

[1] Faculty of Chemical & Natural Resources Engineering, Universiti Malaysia Pahang, Lebuhraya Tun Razak, 26300 Gambang, Kuantan, Pahang, Malaysia
[2] Centre of Excellence for Advanced Research in Fluid Flow, Universiti Malaysia Pahang, 26300 Gambang, Kuantan, Pahang, Malaysia.

[*] Corresponding author. Tel.: +609 549 2874; fax: +609 549 2889.
E-mail: vietvo@ump.edu.my

Abstract. It is the first time ethylene glycol dry reforming (EGDR) reaction was carried out for syngas production on 10%Ni/Al_2O_3 catalyst in a tubular fixed-bed reactor at atmospheric condition. Wet-impregnation method was employed for the synthesis of 10%Ni/Al_2O_3 catalyst. The results of X-ray diffraction measurement revealed the presence of γ-Al_2O_3, NiO, $NiAl_2O_4$, Ni^0 and graphitic carbon phases on the surface of spent 10%Ni/Al_2O_3 catalyst. H_2 temperature-programmed reduction indicates that NiO and $NiAl_2O_4$ phases were totally reduced to metallic nickel phase at reduction temperature above 970 K. EGDR activity appeared to be steady with time-on-stream beyond 5-7 h. Catalytic activity was considerably improved with increasing feed ratio of CO_2:$C_2H_6O_2$ from 1:1 to 2.5:1.

1. Introduction

Syngas referring to the mixture of H_2 and CO has been widely recognized as an economically viable and green energy source to substitute petroleum-based fuels. Typically, this syngas is employed as raw materials in Fischer–Tropsch synthesis (FTS) to produce synthetic liquid fuels and valuable chemicals [1]. Although syngas can be conventionally produced from steam reforming and partial oxidation of methane, high energy consumption due to their endothermic nature and substantial CO_2 greenhouse emission from these industrial processes have induced adversely environmental concerns [2]. Dry reforming of oxygenated hydrocarbons derived from biomass such as ethanol [3] and glycerol [4] has recently gained extensive interests from both academics and industry since this approach not only employs highly available and renewable feedstocks but also mitigates undesirable greenhouse gas discharge by converting it to value-added syngas. However, other types of bio-derived liquids such as ethylene glycol should be also further discovered for syngas generation since it contains the same functional groups present in ethanol and glycerol [5]. In fact, ethylene glycol easily obtained from biomass is also a rich source of hydrogen. Currently, ethylene glycol (EG) aqueous-phase reforming and EG steam reforming have been extensively explored for the production of H_2, but both technologies still contributed to CO_2 emission [6, 7]. Thus, the implementation of dry reforming approach on ethylene glycol could conceivably solve the issue of CO_2 release. Indeed, to the best of our knowledge, research regarding syngas generation from ethylene glycol dry reforming, EGDR (see Eq.(1)) has not been explored before. Based on Gibbs free energy values, it was evident that EGDR reaction was thermodynamically promising at reaction temperature beyond 406 K which is lower than those of ethanol dry reforming (589 K) and methane dry reforming (906 K) [8-10].

$$C_2H_6O_2 + CO_2 \rightarrow 3CO + 2H_2 + H_2O \; (\Delta G_{rxn} = -0.57T_{rxn} + 229.43 \text{ kJ mol}^{-1}) \quad (1)$$

Pt, Pd and Rh based catalysts were reported as outstanding catalysts for syngas production from EG aqueous-phase and steam reforming processes because of their high capacity of C-C bond cleavage, stability and great coking resistance [11-13]. In addition, the employment of those metals during reforming reaction significantly improved the selectivity of H_2. Nevertheless, these noble metals are rarely employed in industrial applications due to limited availability and high cost. Hence, Ni-based catalysts have been considered as favourable catalysts in reforming processes since these non-noble metals are active in cleaving C-C bond and have relatively low cost compared with those precious metals. Thus, the objective of this investigation was to examine the performance of EGDR with time-on-stream (TOS) on 10%Ni/Al_2O_3 catalyst for syngas production.

2. Experimental

2.1. Catalyst synthesis and evaluation

The 10%Ni/Al_2O_3 catalyst was prepared using a conventional impregnation procedure in which $Ni(NO_3)_2.6H_2O$ precursor solution was magnetically mixed with previously pretreated γ-Al_2O_3 support at ambient temperature for 3 h. The resulting solid mixture was subsequently dried for 24 h at 383 K and air-calcined for 5 h at 873 K in flowing air (50 ml min^{-1}). EGDR tests were conducted in a standard stainless steel fixed-bed reactor placed vertically in a divided tubular furnace at temperature range of 873-1023 K and 1 atm. Approximately 0.1 g_{cat} of catalyst sandwiched by quartz wool was initially reduced in situ in 50 ml min^{-1} of 50%H_2/N_2 reducing mixture at 973 K with a 5 K min^{-1} heating rate. Sample was kept at that temperature for 2 h before exposing to gaseous reactants for EGDR reaction. Ethylene glycol precisely pumped by a KellyMed KL-602 syringe pump was vaporized in a pre-heater and mixed with CO_2 reactant and N_2 diluent gas regulated by Alicat mass flow controllers.

The catalytic performances of 10%Ni/Al_2O_3 catalyst in EGDR reaction was calculated in terms of $C_2H_6O_2$ conversion ($X_{C_2H_6O_2}$) and CO_2 conversion ($X_{C_2H_6O_2}$) by using Eqs. (2)-(3). Indeed, the conversion of ethylene glycol, $X_{C_2H_6O_2}$ to gaseous products was calculated based on atom-H balance between the inlet and outlet of the reactor.

$$X_{C_2H_6O_2}\,(\%) = \frac{2F_{H_2,Out} + 4F_{CH_4,Out}}{6F_{C_2H_6O_2,In}} \times 100\% \quad (2)$$

$$X_{CO_2}(\%) = \frac{F_{CO_2,In} - F_{CO_2,Out}}{F_{CO_2,In}} \times 100\% \quad (3)$$

where F_{in} and F_{out} are the inlet and outlet molar flow rates (mol s^{-1}), respectively.

2.2 Catalyst characterization

N_2 physisorption of support and 10%Ni/Al_2O_3 catalyst was performed using Thermo Scientific Surfer unit at temperature of 77 K for investigating Brunauer-Emmett-Teller (BET) surface area, average pore volume and average pore diameter. X-ray diffraction (XRD) measurements of calcined, reduced and spent catalysts were carried out in Rigaku Miniflex II system, in which Cu monochromatic X-ray radiation (wavelength, λ = 1.5418 Å) was employed at 30 kV and 15 mA. The AutoChem II-2920 system was used for H_2 temperature-programmed reduction (H_2-TPR). Sample pre-treated in purging He gas (50 ml min^{-1}) for dehydration at 373 K was exposed to reducing gaseous agent (10%H_2/Ar and 50 ml min^{-1}) from 373 to 1173 K with a ramping rate of 10 K min^{-1}.

3. Results and discussion

3.1. Catalyst characterization

10%Ni/Al_2O_3 catalyst possessed BET surface area of 171.23 m^2 g^{-1}, average pore volume of 0.28 cm^3 g^{-1} and average pore diameter of 61.47 Å. The slightly drop in BET surface compared to γ-Al_2O_3 support (174.13 m^2 g^{-1}) was due to the successful diffusion of NiO nanoparticles into the pore structure of support. XRD measurements of calcined Al_2O_3 support and 10%Ni/Al_2O_3 catalyst (including calcined, reduced and spent samples) are displayed in Figure 1. The diffraction peaks belonging to γ-Al_2O_3 phase were detected at 2θ = 19.45°, 32.30°, 36.99°, 45.69° and 67.20° for all samples. As seen in Figure 1(b), spinel $NiAl_2O_4$ phase was observed at 2θ = 36.99° and 49.38° (JCPDS card No. 10-0339) whilst the characteristic peaks for NiO phase were located at 2θ of 36.99°, 43.60° and 63.08° (JCPDS card No. 47-1049) for calcined 10%Ni/Al_2O_3 catalyst [14]. However, the NiO phase was not noticed for reduced and spent catalysts (see Figures 1(c) and (d), respectively). In fact, metallic Ni^0 phase was identified at 2θ angle of 44.56°, 52.04° and 76.26° (JCPDS card No. 04-0850) for these samples [14]. The presence of Ni^0 phase on catalyst surface was due to the reduction of NiO and $NiAl_2O_4$ forms during in situ H_2 pretreatment before EGDR reaction in agreement with H_2-TPR results (see Figure 2). As shown in Figure 1(d), a high intensity peak centred at 2θ of 26.38° for spent 10%Ni/Al_2O_3 catalyst belonged to graphitic carbon deposition (JCPDS card No. 75-1621) [15].

Figure 2 shows the H_2-TPR profile of 10%Ni/Al_2O_3 catalyst. The first peak (P1) located at 799 K was due to NiO reduction to metallic Ni^0 form ($NiO + H_2 \rightarrow Ni + H_2O$) whilst the high temperature peak (P2) detected at 960 K was assigned to the reduction of $NiAl_2O_4$ phase possessing a strong metal-support interaction ($NiAl_2O_4 + H_2 \rightarrow Ni + Al_2O_3 + H_2O$) [16].

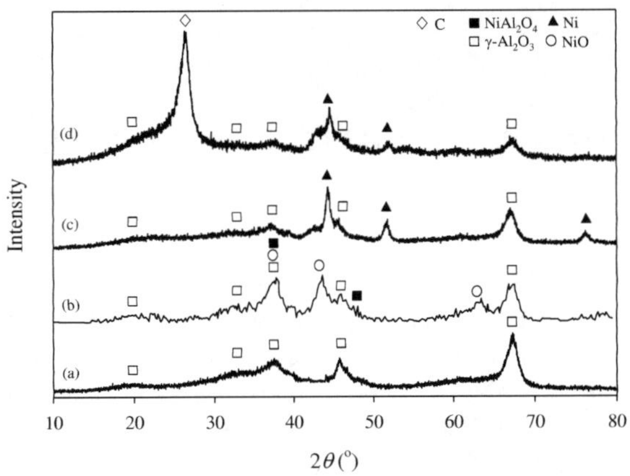

Figure 1. XRD patterns of (a) calcined γ-Al_2O_3 support, (b) calcined 10%Ni/Al_2O_3, (c) reduced 10%Ni/Al_2O_3 and (d) spent 10%Ni/Al_2O_3 catalysts.

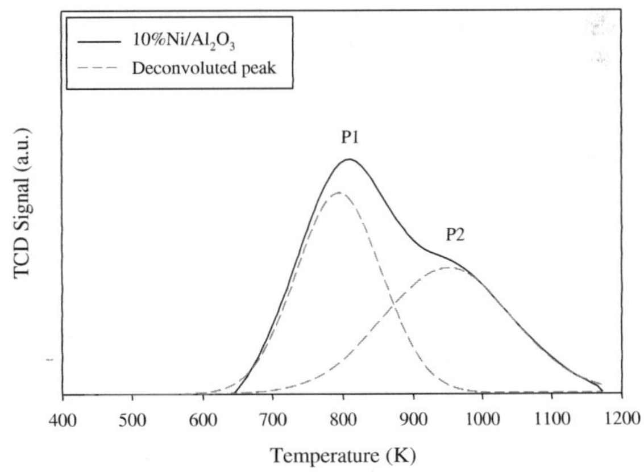

Figure 2. H_2 temperature-programmed reduction profile of 10%Ni/Al_2O_3 catalyst.

3.2. EGDR reaction evaluation

The effect of feed ratio on EGDR performance was determined by adjusting the feed ratios of ethylene glycol and CO_2 at about 2.5:1 and 1:2.5. As seen in Figure 3 (a) and (b), the conversions of ethylene glycol and CO_2 for 10%Ni/Al_2O_3 catalyst improved significantly by up to 84.8% and 66.3%, respectively with the rising of CO_2 feed composition. In the CO_2-rich environment, CH_4 intermediate product generated from ethylene glycol hydrogenation (cf. Eq. (4)) could be further consumed in methane dry reforming reaction (cf. Eq. (5)), in turn enhanced the consumption of both reactants [17, 18]. However, it was evident that only conversion of ethylene glycol was slightly improved with the increment of ethylene glycol whilst a decline in CO_2 conversion. This behaviour could suggest that the adsorption of CO_2 on catalyst surface was suppressed by the existence of ethylene glycol-excess environment. In addition, irrespective of feed ratios, the conversion of ethylene glycol was superior to that of CO_2 conversion. This trend could further verify the presence of parallel side reactions during dry reforming of ethylene glycol, i.e., decomposition, dehydrogenation and steam reforming of EG.

$$C_2H_6O_2 + 3H_2 \rightarrow 2CH_4 + 2H_2O\ (\Delta G_{rxn} = -0.035T - 247.47\ \text{kJ mol}^{-1}) \tag{4}$$

$$CH_4 + CO_2 \rightarrow 2CO + 2H_2\ (\Delta G_{rxn} = -0.28T + 259.04\ \text{kJ mol}^{-1}) \tag{5}$$

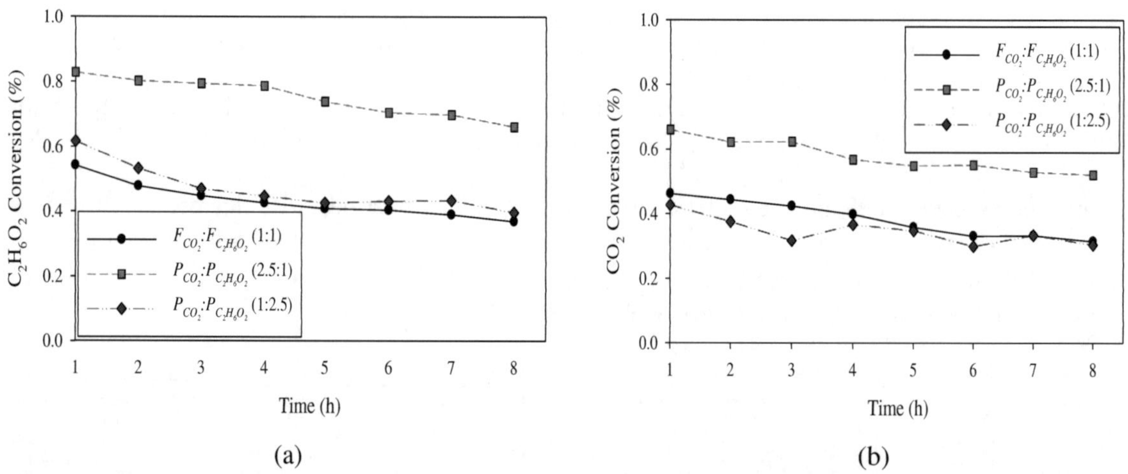

Figure 3. Reactant conversions versus time-on-stream (TOS) during EGDR reaction on 10%Ni/Al_2O_3 catalyst at feed composition of 1:1, 2.5:1 and 1:2.5.

4. Conclusion

The catalytic performance of Ni/Al_2O_3 catalyst prepared by impregnation procedure was successfully investigated in dry reforming of ethylene glycol at different CO_2:$C_2H_6O_2$ ratios of 2.5:1 to 1:2.5 and temperature of 1023 K. XRD measurement revealed the appearance of γ-Al_2O_3, NiO, $NiAl_2O_4$, Ni^0 and carbon phases on the catalyst surface. H_2-TPR revealed that NiO and $NiAl_2O_4$ phases were reduced to metallic Ni^0 phase with in the reduction temperature of 799 and 970 K, respectively. At the stoichiometric condition of EGDR reaction, catalytic activity for 10%Ni/Al_2O_3 catalyst achieved stable condition beyond 5 h on stream. Increasing CO_2 feed ratio enhanced CO_2 and $C_2H_6O_2$ conversions up to 84.8% and 66.3%, respectively owing to the enhancement of methane dry reforming reaction. The decline in CO_2 conversion was experienced with growing ethylene glycol feed composition due to the competing reactant adsorption on catalyst surface in the ethylene glycol-rich environment.

References

[1] Vo D-VN, Arcotumapathy V, Abdullah B and Adesina AA 2013 Non-linear ASF product distribution over alkaline-earth promoted molybdenum carbide catalyst for hydrocarbon synthesis *Catal. Today* **214** 42-49

[2] Abdullah B, Ghani NAA and Vo D-VN 2017 Recent advances in dry reforming of methane over Ni-based catalyst *J. Clean. Prod.* **162** 170-185

[3] Bahari MB, Phuc NHH, Alenazey F, Vu KB, Ainirazali N and Vo D-VN 2017 Catalytic performance of La-Ni/Al_2O_3 catalyst for CO_2 reforming of ethanol *Catal. Today* **291** 67-75

[4] Siew KW, Lee HC, Gimbun J, Chin SY, Khan MR, Taufiq-Yap YH and Cheng CK 2015 Syngas production from glycerol-dry (CO_2) reforming over La-promoted Ni/Al_2O_3 catalyst *Renew. Energy* **74** 441-447

[5] Cesar DV, Santori GF, Pompeo F, Baldanza MA, Henriques CA, Lombardo E, Schmal M, Cornaglia L and Nichio NN 2016 Hydrogen production from ethylene glycol reforming catalyzed by Ni and Ni-Pt hydrotalcite-derived catalysts *Int. J. Hydrogen Energy* **41** 22000-22008

[6] Huber GW, Shabaker JW, Evans ST and Dumesic JA 2006 Aqueous-phase reforming of ethylene glycol over supported Pt and Pd bimetallic catalysts *Appl. Catal. B: Environ.* **62** 226-235

[7] Tupy SA, Chen JG and Vlachos DG 2013 Comparison of ethylene glycol steam reforming over Pt and NiPt catalysts on various supports *Top. Catal.* **56** 1644-1650

[8] Wang W and Wang Y 2009 Dry reforming of ethanol for hydrogen production: Thermodynamic investigation *Int. J. Hydrogen Energy* **34** 5382-5389

[9] Bahari MB, Phuc NHH, Abdullah B, Alenazey F and Vo D-VN 2016 Ethanol dry reforming for syngas production over Ce-promoted Ni/Al_2O_3 catalyst *J. Environ. Chem. Eng.* **4** 4830 - 4838

[10] Usman M, Wan Daud WMA and Abbas HF 2015 Dry reforming of methane: Influence of process parameters-A review *Renew. Sust. Energy Rev.* **45** 710-744

[11] Liu J, Sun B, Hu J, Pei Y, Li H and Qiao M 2010 Aqueous-phase reforming of ethylene glycol to hydrogen on Pd/Fe_3O_4 catalyst prepared by co-precipitation: Metal-support interaction and excellent intrinsic activity *J. Catal.* **274** 287-295

[12] Liu X, Shen K, Wang Y, Wang Y, Guo Y, Guo Y, Yong Z and Lu G 2008 Preparation and catalytic properties of Pt supported Fe-Cr mixed oxide catalysts in the aqueous-phase reforming of ethylene glycol *Catal. Commun.* **9** 2316-2318

[13] Izquierdo U, Wichert M, Kolb G, Barrio VL, Zapf R, Ziogas A, Neuberg S, Arias PL and Cambra JF 2014 Micro reactor hydrogen production from ethylene glycol reforming using Rh catalysts supported on CeO_2 and La_2O_3 promoted α-Al_2O_3 *Int. J. Hydrogen Energy* **39** 5248-5256

[14] Li CY, Zhang HJ and Chen ZQ 2013 Reaction between NiO and Al_2O_3 in NiO/γ-Al_2O_3 catalysts probed by positronium atom *Appl. Surf. Sci.* **266** 17-21

[15] Omoregbe O, Danh HT, Nguyen-Huy C, Setiabudi HD, Abidin SZ, Truong QD and Vo D-VN 2017 Syngas production from methane dry reforming over Ni/SBA-15 catalyst: Effect of operating parameters *Int. J. Hydrogen Energy* **42** 11283-11294

[16] Dieuzeide ML, Iannibelli V, Jobbagy M and Amadeo N 2012 Steam reforming of glycerol over Ni/Mg/γ-Al_2O_3 catalysts. Effect of calcination temperatures *Int. J. Hydrogen Energy* **37** 14926-30

[17] Shabaker JW, Davda RR, Huber GW, Cortright RD and Dumesic JA 2003 Aqueous-phase reforming of methanol and ethylene glycol over alumina-supported platinum catalyst *J. Catal.* **215** 344-352

[18] Shabaker JW, Simonetti DA, Cortright RD and Dumesic JA 2005 Sn-modified Ni catalysts for aqueous-phase reforming: Characterization and deactivation studies *J. Catal.* **231** 67-76

Acknowledgments

The authors are grateful for the financial support from Universiti Malaysia Pahang (UMP) via UMP Research Grant Scheme (RDU170326) to conduct this research.

IOP Conf. Series: Materials Science and Engineering **446** (2018) 012014 doi:10.1088/1757-899X/446/1/012014

Research on Optimal Operation of AC/DC Active Distribution Network Based on SOP

Zhanjun LI[a], Zhikai WU[a], Li JIANG[a], Heyan ZHU[a],Ming MENG[b],Yi WEI[b], Guolin ZHU[b]

a State Grid Liaoning Electric Power Company Limited Economic Research Institute, Shenyang 110015,China
[b]North China Electric Power University, Baoding 071003, China

Email: mmwxp@126.com

Abstract. The development of AC/DC active distribution network promotes the improvement of distribution network power supply mode. As a new-type power electronic device,Soft Open Point(SOP) has many benefits for the grid operation. In this paper, the AC/DC active distribution grid was constructed firstly and the control strategy of the back-to-back voltage source of SOP was analyzed in detail. Secondly, a mathematical model was established to reduce the system loss and to improve the node voltage over-limit condition. Considering these problems can be taken as multi-objective and nonlinear optimize problems, this paper put forward the improved genetic algorithm to solve it. Finally, the results of the simulation indicates the effectiveness of the proposed algorithm.

1. Introduction

The development of new energy technologies and the access of a large number of distributed power sources have led to a decrease in the operating efficiency of traditional distribution networks, an increase of system costs, and the reduction operational reliability[1]. The AC/DC active distribution network can autonomously adjust the system voltage and power flow according to the actual operating state of the power system, and realize the coordinated control and optimal operation of the source-network-load level to meet the operational metrics of the power system economy and safety. It is one of the most important forms of active distribution networks in the future [2].

The rapid development of new energy technologies makes AC distribution networks increasingly being unable to meet the needs of the current society. Photovoltaic power generation and storage battery access to AC distribution networks require DC/DC converters. Wind power generation access to AC distribution networks also requires AC/DC/AC converters [3]. The AC line containing several new energy devices will be converted into a DC line, and an AC/DC active distribution network will be constructed in order to meet the safe operation of the distribution network under the premise of reducing converter device, reducing the system cost and thus is the feasible scheme of power distribution network frame optimization[4]. SOP can replace the contact switch in the traditional AC line, flexibly control the connected lines, and can improve the current control status of the "Closed-loop Design, Open-loop Operation" of the traditional distribution network..

Under this circumstance, this paper proposes an active power distribution network frame structure based on SOP in order to reduce power loss of network and to improve the voltage over-limit condition firstly. On this basis, a mathematical model was set up for the purpose of optimization. It is necessary to consider multiple output variables in the operation optimization of AC/DC active distribution network with SOP which belongs to the optimization problem of mixed-integer nonlinear function[5]. In this paper, an improved genetic algorithm is used to solve the mathematical model. The improved genetic algorithm is faster than the traditional genetic algorithm, and it is not easy to be trapped in the local solution. Finally, a simulation model was built in Matlab/Simulink to verify the correctness of the AC/DC active distribution network structure and the effectiveness of the proposed algorithm.

2. AC / DC active grid structure with SOP

The structure of the traditional AC distribution network system is presented, as shown in Fig.1(a). The power distribution system consists of two 10kV medium-voltage feeders stepped down by a transformer. The connection switch connects them. Each AC line is connected to a photovoltaic device and load. The contact switch in Line 1 can realize the optimal operation of the traditional distribution network. This paper selects the SOP for power electronics as a substitute for some contact switches, transforms AC line with multiple distributed power sources into DC line, and uses inverter VSC connected DC line with AC line to build a hybrid AC/DC distribution network. The transformed AC/DC active distribution network system structure is proposed, as shown in Fig.1(b). The line 2 contains several photovoltaic devices which is converted into a DC line. The VSC is placed at the connection of the first end of the line 2 and the two lines. The SOP in place of the contact switch in line 1.Based on the AC/DC active distribution network composed of power electronic devices, the coordination between SOP and VSC is adopted to realize the power flow control between feeders and the optimal operation of the system.

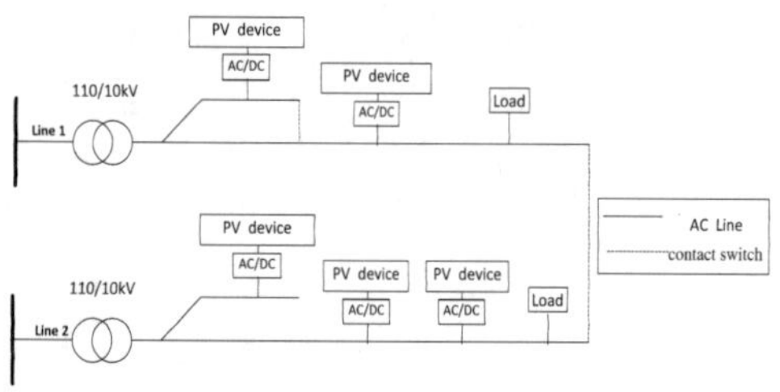

(a)

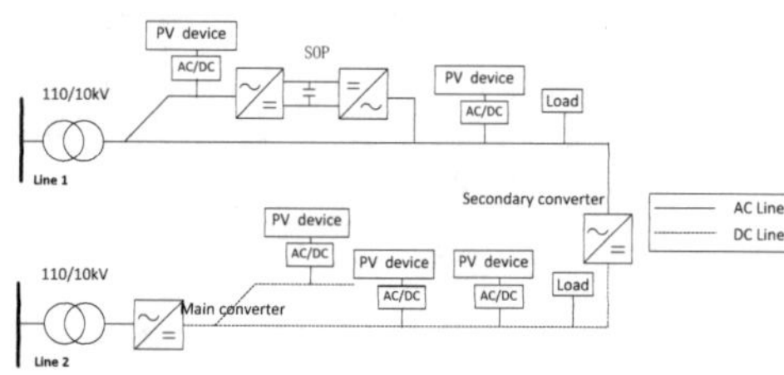

(b)

Figure 1. AC/DC active distribution network system structure:(a) Before remodeling (b)After remodeling

3. Optimal modeling and solution of AC/DC active distribution network with SOP

3.1. *Control system of SOP*

Currently, there are three general structures of SOP: back-to-back voltage source converter (B2B VSC), static synchronous series compensator (SSSC), and unified power flow controller (UPFC) [6]. This paper takes back-to-back voltage source converter as an example to introduce the control strategy of SOP in AC/DC active distribution network.

B2B VSC system consists of two parts of four-quadrant VSC1 and VSC2, DC filter capacitor C, VSC converter grid-side equivalent connection resistance R1, R2 and inductance L1, L2. The filter capacitor C is used to provide DC voltage support and reduce the ripple of the DC bus voltage. L1 and L2 are used to filter out the harmonics of the converter output current. R1 and R2 represent the equivalent resistance of losses between the converter and the AC system.

The goal of this paper is to achieve coordinated control of SOP and VSC in order to reduce system network losses and to improve voltage over-limitation. The AC side voltage of the main converter station is determined by the 10kV bus voltage and the DC side voltage level is 15kV. The control method is Vdc-Q. AC and DC lines are connected from the secondary converter station to accurately adjust the active power and reactive power. The control method is P-Q[7]. SOP outer ring needs to output the AC side voltage, active power, and reactive power in real time according to the current operating state of the system so as to achieve effective control of the system. At this time, the VSC on both sides need to coordinate with each other. In summary, this paper selects the SOP control strategy as PQ-VdcQ control, which is one converter controls the DC side voltage, and the other converter controls the transmission power.

3.2. Mathematical model

Considering that a large number of distributed power sources complicate the operation behavior of distribution networks, traditional static power flow cannot satisfy the optimization requirements. Thus a dynamic power flow optimization method was adopted to reduce power loss of network and to improve voltage over-limit conditions as an objective function. To optimize the problem, this paper establish a mathematical model using the weight coefficient transformation method, so as to converts the multiple objectives into a single goal. Ignoring the power loss of the SOP, the optimized objective function can be expressed as:

$$\min F = C_1(P_{loss}^{AC} + P_{loss}^{DC} + P_{loss}^{VSC}) + C_2(\sum_{i=1}^{n}|V_i - V_N|) \tag{1}$$

In the formula, C_1 and C_2 represent the weight coefficient, and assumed C_1=C_2=0.5. P_{loss}^{AC} indicates loss of AC line, P_{loss}^{DC} indicates loss of DC line, P_{loss}^{VSC} indicates loss of inverter, V_i indicates voltage value of node *i*, and V_n indicates the rated voltage.

The operational constraints of the distribution network include active and reactive power system flow constraints, converter VSC constraints, SOP operating constraints, node voltage upper and lower limit constraints.

$$P_i = U_i \sum_{j=1}^{N} U_j (G_{ij}\cos\theta_{ij} + B_{ij}\sin\theta_{ij}) \tag{2}$$

$$Q_i = U_i \sum_{j=1}^{N} U_j (G_{ij}\sin\theta_{ij} - B_{ij}\cos\theta_{ij}) \tag{3}$$

$$U_{i\min} \le U_i \le U_{i\max} \tag{4}$$

$$\sqrt{P_{VSC}^2 + Q_{VSC}^2} \le S_{\max}^{VSC} \tag{5}$$

$$\sqrt{P_{SOP1}^2 + Q_{SOP1}^2} \le S_1^{SOP} \tag{6}$$

$$\sqrt{P_{SOP2}^2 + Q_{SOP2}^2} \le S_2^{SOP} \tag{7}$$

$$P_{SOP1} + P_{SOP2} = 0 \tag{8}$$

In the formula, P_i and Q_i represent the active power and reactive power respectively injected into node *i*. G_{ij} and B_{ij} are the real part and the imaginary part of the admittance matrix of the system. θ_{ij} is the phase angle difference between nodes *i* and *j*. U_{imin} and U_{imax} indicate the upper and lower limits of the node voltage *i*. P_{vsc} and Q_{vsc} indicate the active power and reactive power flowing through the converter. $S_{\max}^{VSC}$ indicates the upper limit of the converter capacity. $Psop1$、P_{sop2} indicate the active power flowing through the SOP; Q_{sop1}、Q_{sop2} indicate the reactive power flowing through the SOP; S^{sop} indicates the operating capacity of the SOP.

3.3. Solving algorithm

For the optimization model proposed in the previous section, this paper uses an improved genetic algorithm to solve it. When solving the optimization problem of AC/DC active distribution network based on SOP proposed in this paper, the genetic algorithm first encodes and initializes the control variables in the AC/DC system, then brings individuals into the power flow calculation equation and calculates their adaptive value functions; The convergence solution obtained through selection, crossover, and mutation is the optimal power flow for each individual, and then it is taken into the power flow calculation equation to calculate its fitness; The individual objective function values are sorted, and the bad individuals are replaced by excellent individuals, the next round of optimization is performed until an optimal solution is obtained or the upper limit of the number of iterations is reached. This method can effectively avoid falling into a local optimum when solving optimization problems.

3.4. .Analysis of examples

Using Matlab/Simulink to build the AC/DC active distribution network model to verify the correctness of the optimization algorithm. The AC bus voltage is 10kV;The DC bus voltage is 15kV; the rated capacity of the photovoltaic power station is 2MW, and the transmission efficiency is 0.95; the rated capacity of the VSC converter station is 5MW, and the transmission efficiency is 0.98; the rated capacity of the SOP is 5MW, and the transmission efficiency is 1; AC power factor reference is 1.

Fig2(a) shows the power loss generated by the AC/DC active distribution network with and without using the optimization algorithm.The system loss without optimization algorithm is shown as curve 1. The system loss mainly includes the line loss caused by AC line and DC line, VSC converter station loss, etc. The total loss is about 0.5 MW; when the system adopts the optimization algorithm proposed in this paper , the loss of the system is significantly reduced by coordinating the SOP, VSC converter station and photovoltaic power generation unit. After optimization, the loss is about 0.2 MW, as shown in curve 2 of the figure a.

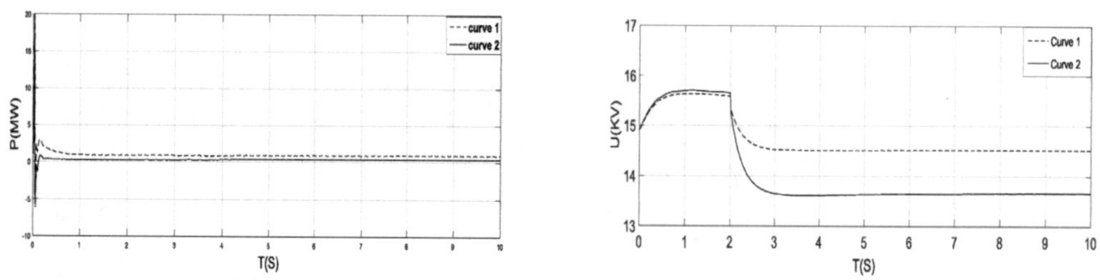

(a) (b)

Figure 2. With and without using the optimization algorithm system comparison chart: (a) system losses (b) the node voltage

Fig2(b) shows the results of using of an optimization algorithm to optimize the voltage at a certain point on the DC side of the AC/DC active distribution network. In actual operation, the unstable output of the distributed power device and the sudden change of the load value will generate voltage fluctuations in the grid voltage and cause the node voltage to exceed the limit. Curve 2 indicates that the load on the DC side increases at 2 seconds. It can be seen that the voltage suddenly drops to about 13.75 kV when the node voltage has crossed the line. After the optimization algorithm was adopted, the lower limit of the voltage was significantly increased, which was about 14.5kV. The voltage of the node did not cross the line, as shown in curve 2 of the Fig2(b).Figure2 shows that the optimized control algorithm proposed in this paper can effectively reduce the system losses and improve the voltage over-limit conditions caused by the unstable output of the distributed power supply by coordinating and controlling the converter station and SOP.At the same time, it also proves that the correctness of the grid.

4. Conclusion

The rapid development of DC-type distributed power supplies, energy storage devices, and electric vehicles has made DC power distribution more and more became a hot topic of research. SOP is a power electronic device that can be used to replace contact switch in traditional power grids and has a significant potential in application, and may even create a new distribution network power supply model.From the perspective of economic operation, this paper established a distribution network optimization operation model, adopted an improved genetic algorithm to solve the mathematical model, and the simulation results verified the rationality of the AC/DC active distribution network structure and the correctness of the proposed optimization algorithm. This method is effective to reduce energy consumption, to improve the voltage over-limit condition, and meets the requirements of the economical operation and green development of the active distribution network.

References

[1] Romero-Ramos E, Marano-Marcolini A. Assessing the loadability of active distribution networks in the presence of DC controllable links. IET generation, Transmission&Distribution,2011;5(11):1105.
[2] S K Chaudhary, Guerrero J M. Enhancing the capacity of the AC distribution system using DC interlinks-a step toward future DC grid.IEEE Transactions on Smart Grid,2015;6(4):1722-1729.
[3] S.K.Chaudhary, E.Demirok, R.Teodorescu. Distribution system augmented by DC links for increasing the hosting capacity of PV generation.In:Proc.Power.Electron.Drives.Energy Syst.(PEDES),2012:1-5.
[4] L.Xiong, W.Peng, P.Loh. A hybrid AC/DC microgrid and its coordination control.IEEE Trans.Smart Grid,2011;2(2):278-286.
[5] C.Chen, C.Tsai,C.Lin,etal.Loading balance of distribution feeders with loop power controllers considering photovoltaic generation.IEEE Trans.Power Syst.2011;8(26):1762-1768.
[6] Wangyu Cao, Nick Jenkins, Jianzhou Wu. Feeder Load Balancing in MV Distribution Networks Using Soft Normally-Open Points. In:2014 5th IEEE PES Innovative Smart Grid Technologies Europe(ISGT Europe),2014:1-6.
[7] Alcala J, Cardenas V, Ramirez-Lopez A R,et al.Study of the bidirectional power flow in back-to-back converters by using linear and nonlinear control strategies.Presented at:2011 IEEE ECCE.

IOP Conf. Series: Materials Science and Engineering **446** (2018) 012015 doi:10.1088/1757-899X/446/1/012015

Research on Optimal Control of AC-DC Hybrid Distribution Network

Zhikai WU[a], Dalong HU[a], Jing BAI[a], Li JIANG[a], Ming MENG[b], Guolin ZHU[b], Yi WEI[b]

[a]State Grid Liaoning Electric Power Company Limited Economic Research Institute, Shenyang 110015,China
[b]North China Electric Power University, Baoding 071003, China

Email: mmwxp@126.com

Abstract. It is hard to optimize power flow and operating status of hybrid AC/DC distribution network with traditional control strategy, so this paper proposed a hierarchical control strategy with different voltage scales. In the first layer, different converter stations maintain stable operation of power system with different control strategy; In the second layer, when power fluctuations of system are large, converter stations change its operation mode to balance voltage fluctuation of DC side; In the third layer, energy optimal dispatch system calculate different converter stations' control reference to reduce power loss of system. When operation of system change, systems' stable operation is maintained by switching of control strategy in different layer. The simulation using Matlab/Simulink shows that this control strategy can reduce power loss and system can operate stably even in communication failure.

1. Introduction

With social progress and economic development, power load is growing rapidly, distributed generation is being paid more and more attention, besides, the user's power consumption has also changed greatly, the proportion of DC load is increasing largely. With the change of power supply mode and power consume mode, AC distribution network meet great challenge in access to distributed power and improve the power quality, so hybrid AC/DC distribution network is being paid more and more attention by experts and scholars [1]-[4].

To improve the reality of distribution network, experts try to connect different AC distribution network by flexible interconnected device(FID) and DC line, every area of the system can support power to each other by DC line. Besides, distribution generation, DC load and energy storage system can be connected to DC line. For the structure of hybrid AC/DC distribution network is much more complex, how to coordinate the control strategy of different converter stations is the key of system's stable operation.

In view of the above problems. A kind of control strategy was proposed to control AC voltage and frequency by converter station and energy storage system maintain the DC voltage within the allowable range by droop control, but this control strategy did not consider overcharge and discharge of energy storage systems [5]. Aim at the fluctuation of DC line voltage of hybrid AC/DC distribution network, feedforward controller and anti disturbance controller was adapted to balance it [6]. To reduce the impact of branch voltage fluctuation caused by distribution generation and electric vehicle, the VSC of the branch which photovoltaic generation unit was connected into AC distribution network adapt AC voltage-power control [7]. A kind of control strategy of hybrid AC/DC distribution network based on back to back interconnected converter was proposed, it gives different converters control strategy depend on the power transmission between DC network and AC network, it also consider many kind of failure condition at the same time [8].

To optimize systems' operating status without need excessive communication requirements, this paper combine optimal scheduling with traditional control. When system operating normally, optimal scheduling system provide converter station with dispatch instruction to reduce its power loss; when dispatching system work abnormally, it change converters' mode to stable distribution network by detecting the voltage of each node.

2. Hybrid AC/DC Distribution Network's Structure

Nowadays, the development of AC and DC hybrid distribution network is in the initial stage, its topology is usually referred to traditional distribution network, this paper focuses on the topological structure of two-terminal. Connecting AC distribution network to DC distribution network by flexible interconnected device, energy storage system, photovoltaic power generation system and AC load are connected into DC areas by different flexible interconnected device, DC load and double fed induction generator(DFIG) are connected into AC areas. It can reduce the use of converter stations when more electric unit need to connected into it, besides, power support between AC areas and DC areas can be realized through effective control of flexible interconnect devices. Its topology shown in Fig. 1.

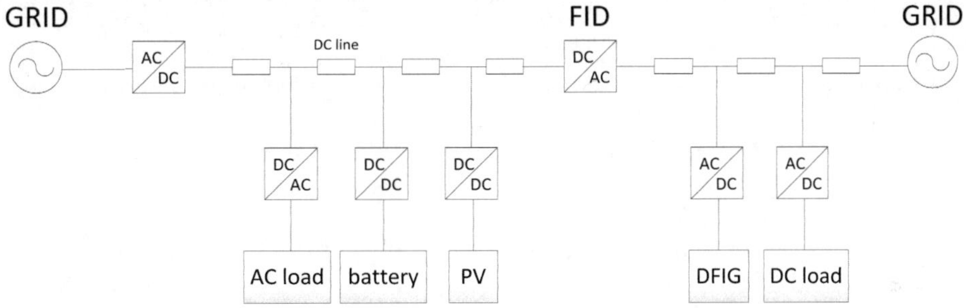

Figure 1. Topology of hybrid AC/DC distribution network.

3. Control Strategy

The hierarchical control proposed in this paper is divided into three layers according to DC side voltage. when the DC voltage deviation is within the allowable range, the main converter station and the minor converter station adopt different control strategies to provide voltage support for the DC side at the same time; when the DC voltage deviation become larger, energy storage system adopt droop control strategy to maintain the DC voltage; when systems operate stably, optimal scheduling system provide scheduling instruction for converter stations. Control strategy of each layers as follow.

3.1. The first layer's control strategy

In the first layer, DC voltage deviation is small, the main converter adopt constant voltage control strategy, the minor converter adopt U-P droop control strategy, both of it support power for DC areas, besides the control parameters are provided by optimal power flow calculation. The voltage can be expressed as (1) and (2).

$$U - U_{opf} = -k_{up}(P_{VSC2} - P_{opf}) \tag{1}$$

$$k_{up} = \min\left\{\frac{U_{\max} - U_{opf}}{P_{opf} - P_{\min}}, \frac{U_{opf} - U_{\min}}{P_{\max} - P_{opf}}\right\} \tag{2}$$

where U_{opf} and P_{opf} are the control parameter reference, $U_{\max}$, $U_{\min}$, $P_{\max}$, $P_{\min}$ are critical point of mode switching, k_{up} is its drooping coefficient.

3.2. The second layer's control strategy

When the system is in poor operating condition, converter stations' voltage and power may exceed its critical value and change its control mode automatically, it is hard to maintain DC voltage in allowable range with only the main converter and minor converter. The energy storage systems will switch to droop control at this time to recover voltage to normal status. The voltage can be expressed as (3).

$$U_{Bi} = U^*_{Bi} - k_{Bi}(P_{Bi} - P_{Biref}) \tag{3}$$

where P_{Bi} and P_{Biref} are power and its reference of battery, U^*_{Bi} is voltage's critical value, k_{Bi} is its drooping coefficient.

3.3. The third layer's control strategy

The third layer is energy optimization dispatch system, transfer the date collected or predicted to the optimal power flow calculation system, calculate and get the reference of control which will let the system in the optimal status. When the third layer work normally, it offers its calculation results to the lower layer.

In summary, the control strategy of this paper can be shown in Fig. 2

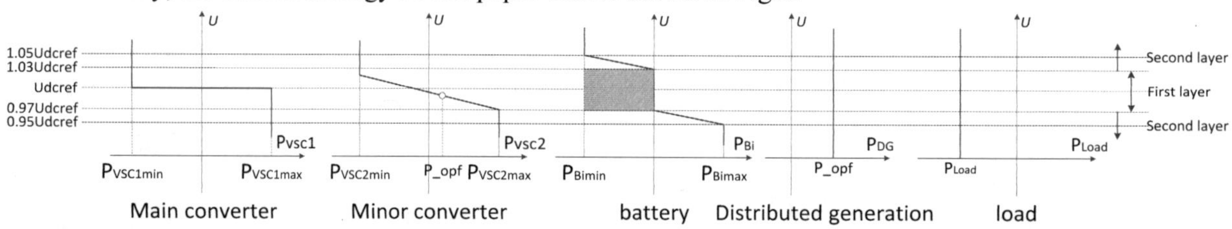

Figure 2. Proposed hierarchical control strategy

4. Energy Optimization Scheduling Model

The target of this control is to minimize the power loss of hybrid AC/DC distribution network and satisfy the constraint conditions at the same time, so objection function of system should include three part, minimize the power loss of line and converter and maximize distributed generation's power, so its objective function can be expressed as (4).

$$f_1(u,x) = \sum_{i=1}^{n_{DG}} (P_{DGi\_MPPT} - P_{DGi}) + \sum_{i=1}^{n_{VSC}} P_{VSCi}(1-\eta_i) + P_{loss\_line} \tag{4}$$

Where P_{DGi} and P_{DGi_MPPT} are actual power and maximum power generation power of distribution generation; P_{VSCi} is actual power of converter station, η_i is its efficiency; P_{loss_line} is the power loss of DC and AC transmission line

When hybrid AC/DC distribution network work stably, it should satisfy equality constraint and inequality constraint which can be expressed as follow formula

4.1. constraint of node power

$$P^{DC}_{DGi} + P^{DC}_{VSCi} - P^{DC}_{Li} - U_i \sum_{j=1}^{N} (U_i - U_j) g_{ij} = 0 \tag{5}$$

$$P^{AC}_{DGi} + P^{AC}_{VSCi} - P^{AC}_{li} - U_i \sum_{j=1}^{N} U_j [G_{ij}\cos(\theta_i - \theta_j) + B_{ij}\sin(\theta_i - \theta_j)] = 0 \tag{6}$$

$$Q^{AC}_{DGi} + Q^{AC}_{VSCi} - Q^{AC}_{li} - U_i \sum_{j=1}^{N} U_j [G_{ij}\sin(\theta_i - \theta_j) - B_{ij}\cos(\theta_i - \theta_j)] = 0 \tag{7}$$

Where DG_i, VSC_i and L_i are distribution generation, converter station and load. DC and AC mean DC areas and AC areas.

4.2. constraint of node voltage

$$U_{i\min} \le U_i \le U_{i\max} \tag{8}$$

where U_i, $U_{i\min}$ and $U_{i\max}$ are node voltage and its limit.

4.3. constraint of transmission line power

$$P_{ij\min}^{DC} \le P_{ij}^{DC} \le P_{ij\max}^{DC} \tag{9}$$

$$S_{ij\min}^{AC} \le S_{ij}^{AC} \le S_{ij\max}^{AC} \tag{10}$$

where P_{ij}, $P_{ij\min}$ and $P_{ij\max}$ are actual power of transmission line and its limit; S_{ij}, $S_{ij\min}$ and $S_{ij\max}$ are apparent power of transmission line and its limit.

4.4. constraint of converter station

$$P_{VSCi\min} < P_{VSCi} < P_{VSCi\max} \tag{11}$$

$$\sqrt{P_{VSCi}^2 + Q_{VSCi}^2} < S_{VSCi} \tag{12}$$

where P_{VSCi} and Q_{VSCi} are actual power and reactive power of converter station, $P_{VSCi\min}$, $P_{VSCi\max}$ and S_{VSCi} are its limit.

5. Simulation Analysis

To verify the effectiveness of the hierarchical control strategy in this paper, building the simulation model shown in Fig. 1 with Matlab/Simulink, where the DC bus voltage is 20Kv, DC voltage bus voltage is 6Kv. The following gives the system operating status when power fluctuate, its operating conditions change as follows and the status of system shown in Fig. 3

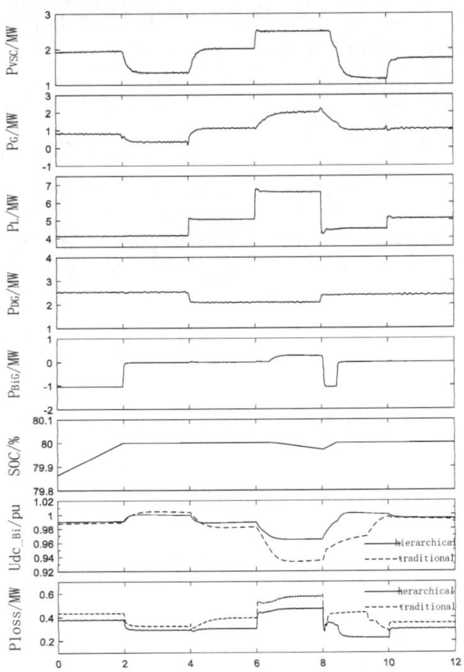

Figure 3. System operation performance

In Initial state, power of PV and DFIG are 1.5Mw and 1.1Mw, load L1, L2 and L3 are 2Mw, 1Mw and 1Mw, state of charge of energy storage system is 79.86%, its voltage deviation is small; At the 2s, SOC reach 80%, battery stop charging; At the 4s, energy scheduling system hitch, besides, PV's power reduce to 1Mw, load L2 increase to 2Mw, it lead to drastic decrease of DC voltage; At the 6s, load L1 increase to 3.5Mw, the main converter station's power reach its upper limit and switch its control strategy to current limiting mode which will lead to drastic decrease of DC voltage. when the DC voltage deviation reach its lower limit, the systems will switch to the second layer. At the 8s, load L1 reduce to 1Mw, load L3 increase to 1.5Mw and power of DFIG increase to 1.5Mw, the demand of DC load is reduced at that time, so the DC voltage increase; At the 10s, the scheduling system return to normal.

6. Conclusion

This paper presents the hybrid AC/DC distribution network's structure and its hierarchical control strategy, establish an optimal scheduling model which aims at minimal power loss. It can be known through simulation verification that this paper's control strategy can reduce power loss and ensure the stable operation of the system with no need for high communication quality; when system's power fluctuates, minor converter change its power under droop control, reduce the pressure of the main converter station.

References

[1] Bracale A, Cambia P, Carpinelli G, et al. A hybrid AC/DC smart grid to improve power quality and reliability[J]. Energy Conference and Exhibilition, 2012: 507-514.

[2] Zeng Jiasi, Xu Xidong, Zhao Yuming. Reliability comparison of AC and DC Distribution Network[J]. Power System Technology, 2014, 38(9): 2582-2589.

[3] Radwan A,Mohamd Y. Assessment and mitigation of interaction dynamics in hybrid AC/DC distribution generation system[J]. IEEE Transaction on Smart Grid, 2012, 3(3): 1382-1393.

[4] Pei Wei, Deng Wei, Zhang Xue, et al. Potential of using multi-terminal LVDC to improve plug-in electric vehicle integration in an existing distribution network[J]. IEEE Transactions on Industrial Electronics, 2015, 62(5): 3101-3111.

[5] Kyohei K, Jomonbu S, Atshushi Y, et al. A hybrid smart AC/DC power system[J]. IEEE Transaction on Smart Grid, 2010,1(2): 199-204.

[6] Masoud D, Yasser A I. Robust multi- objective control of VSC-based DC-voltage power port in hybrid AC/DC multiterminal micro-grids[J]. IEEE Transaction on Smart Grid, 2013, 4(3): 1597-1612.

[7] Sanjay K C, Iosep M G, Remus T. Enhancing the capacity of the AC distribution system using DC interlinks-a step toward future DC grid[J]. IEEE Transaction on Smart Grid, 2015, 6(4): 1722-1729.

[8] Ritwik M. A Hybrid microgrid with DC connection at back to back converters[J]. IEEE Transactions on Smart Grid, 2014, 5(1): 251-259.